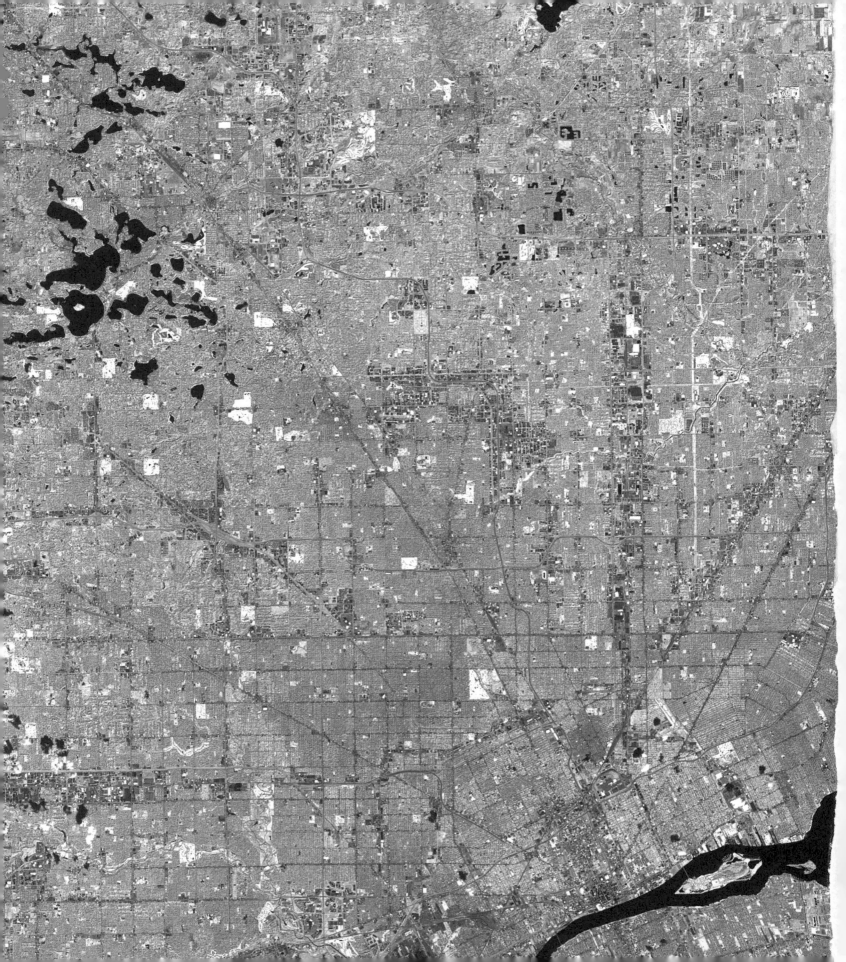

City Unseen

City Unseen
New Visions of an Urban Planet

Karen C. Seto and Meredith Reba

Yale UNIVERSITY PRESS

New Haven and London

Published with assistance from the Alfred P. Sloan Foundation.

Yale University Press books may be purchased in quantity for educational, business, or promotional use. For information, please e-mail sales.press@yale.edu (U.S. office) or sales@yaleup.co.uk (U.K. office).

Selected illustrations by Barbara Schoeberl (electromagnetic spectrum and color images in the "Views from Space" chapter).

Typesetting and design by Lindsey Voskowsky.

Cartography and design by Santiago Cortes Villota.

Printed in China.

Library of Congress Control Number: 2017953535
ISBN 978-0-300-22169-5 (hardcover : alk. paper)

A catalogue record for this book is available from the British Library.

This paper meets the requirements of ANSI/NISO Z39.48-1992 (Permanence of Paper).

10 9 8 7 6 5 4 3 2 1

Frontispiece: Guayaquil, Ecuador (in light blue)

For our mothers

contents

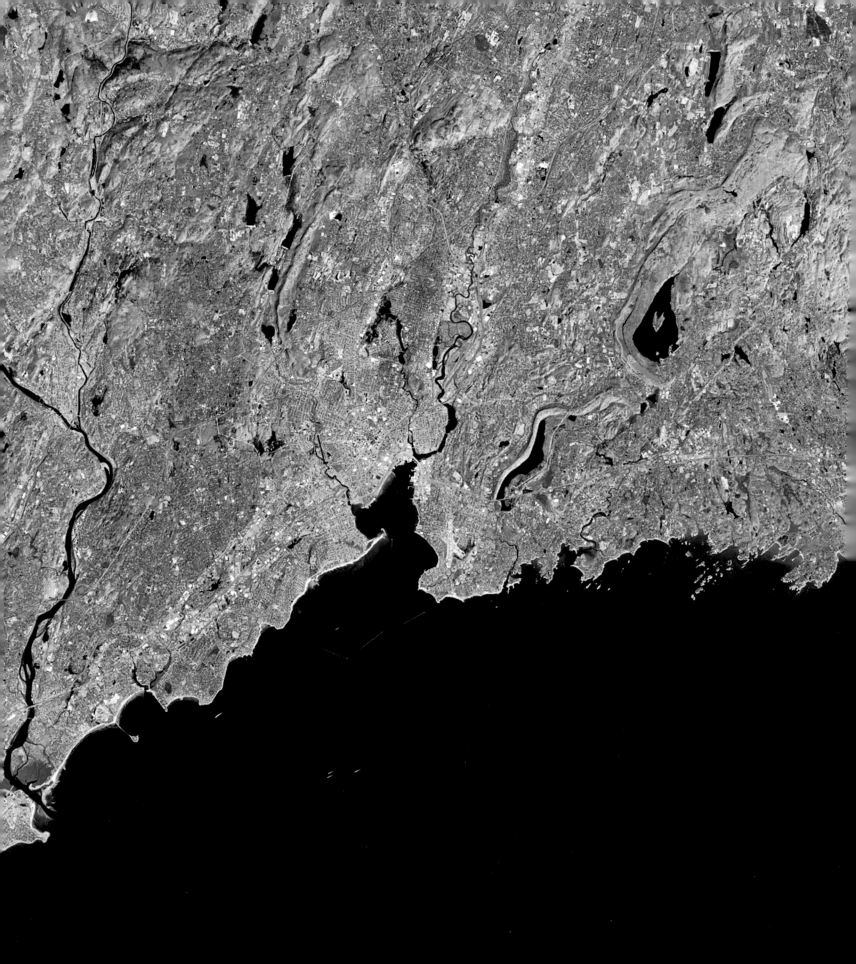

foreword

You may have visited the cities that appear in this book. You may even live in one. And yet many of these images show scenes unlike anything that you or I have ever seen. These images present the city at an unusual scale—one that highlights how humanity and the planet work together or in opposition. We see urban areas shining in unexpected colors—casting even the most familiar places in an almost otherworldly light to reveal a better understanding of vegetation, water, and human settlements. This isn't the first time a change of perspective has made familiar places on Earth suddenly seem foreign and reshaped my understanding of our home planet.

As a space shuttle astronaut, I experienced the Earth as a beach ball suspended against the velvety black, star-studded backdrop of space. Zooming around it at 17,500 miles per hour turned my earthbound sense of time and distance completely upside down. At that speed, a full lap around the planet takes just ninety minutes. No place on Earth is more than minutes away, the sun rises or sets every forty-five minutes, and familiar time zones are meaningless. We watched the incredibly sharp line separating day and night race across the landscape, flew through the iridescent green curtains of the aurora, and saw shooting stars streak through the atmosphere below us. Patterns and features I knew well thanks to my lifelong love of maps seemed to snap into

clearer focus as my new perspective revealed them to me in a vastly wider, richer context. The images in *City Unseen* struck me in much the same way as my views out the spacecraft window: stunning, beautiful, and enlightening, while also intriguing and sometimes disorienting.

One goal on my first spaceflight was to launch the Earth Radiation Budget Satellite (ERBS). Much like the satellites that recorded the images in this book, ERBS monitored the ways our planet reflects the energy that the sun emits. And much like the data used to create this book's urban images, data from that satellite helped us to investigate and improve our relationship to the environment. Scientists used ERBS data to understand how chlorofluorocarbons (CFCs), emitted during the burning of fossil fuels, depleted the ozone. Better information led to action. The signing of an international agreement in 1987 (the Montreal Protocol) led to major reductions in CFC emissions and the restoration of the ozone layer—a tremendous environmental success story.

Despite the historic role of ERBS, it is not the most famous satellite mission of my career. On my second spaceflight, in April 1990, I was a member of the five-day space shuttle *Discovery* mission that launched the Hubble Space Telescope. Those of you who have seen the menagerie of galaxies recorded in Hubble

New Haven, Connecticut, United States (center, in light blue)

images understand that this satellite looks not at Earth but out into space, recording light that has traveled for long periods of time, up to 13.2 billion years, allowing us to peer into the universe's farthest reaches and distant past.

To paraphrase my former colleague John Glenn, helping humanity look outward does not prevent it from looking homeward. My work at the National Aeronautics and Space Administration (NASA) led to a career at the National Oceanic and Atmospheric Administration (NOAA), where I asked many of the same questions as this book's authors. In particular, my goal at NOAA was to understand and predict changes in our ocean and atmosphere, making this scientific information useful to society. From daily weather forecasts to tsunami and hurricane warnings, drought outlooks, and nautical charts, NOAA information products offer everyone from heads of house-hold to heads of state a better understanding of how changing conditions may affect our planet and the humans who live here, allowing them—we hope—to make better decisions.

I hope the perspectives presented in this book also help you to look outward and inward, to understand the global world in which we live, and to find it in yourself to ask deep questions about our urban futures.

Kathryn D. Sullivan, oceanographer, retired NASA astronaut, and former administrator of the National Oceanic and Atmospheric Administration

North and South America

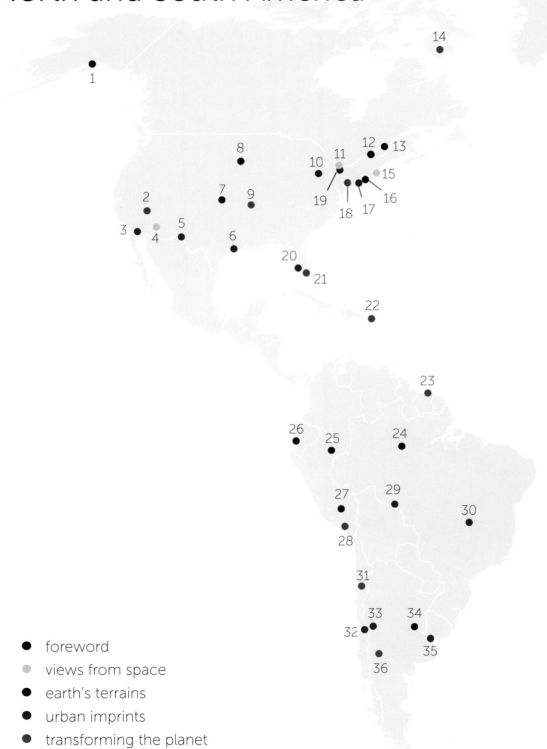

- foreword
- views from space
- earth's terrains
- urban imprints
- transforming the planet

Europe and Africa

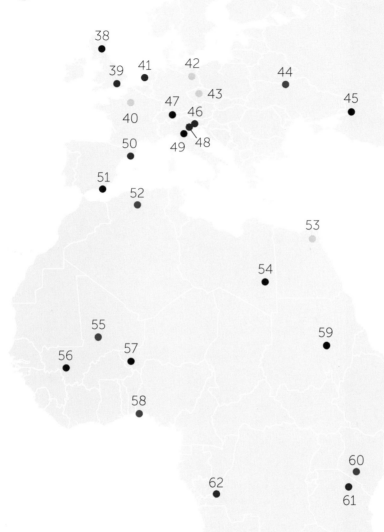

37

38

39   41   42   44

47   43   45

40   46

50   49   48

51

52

53

54

55

57   59

56

58

60

62   61

- ● views from space
- ● earth's terrains
- ● urban imprints
- ● transforming the planet

Asia, Oceania, and Antarctica

views from space

earth's terrains

urban imprints

transforming the planet

Key

North America

South America

Europe

Oceania

Antartica

introduction

The train lurched out of the station, and we chugged our way through the narrow streets. My window exposed glimpses of morning routines: a baby being fed; a man brushing his teeth; an elderly woman stirring food in a wok; a child patting down a school uniform; a young girl walking to work. It was like watching many movies through a single frame. Soon the train left the city behind, and we moved through open lands checkered with flooded rice paddies, fish farms, and bright vegetables planted in rows. The transition from city to countryside was gradual but evident: from dense streets, alive with human activity, to sparse landscapes of verdant parcels with few people and fewer actions, from the cacophony of bowls, eating utensils, and human voices to only the uninterrupted hissing, rumbling, and chugging of the train. From the train window, the remaining urban imprint softened, and our view was one shimmering fertile field after another. A building here and there; a cluster of low-rise brick homes; a child on a bicycle; people hunched over, harvesting crops from the root. Within minutes, we went from the city to the larger world. How could the city be erased so quickly, and yet still be so near?

The age and location of the world's oldest city are up for debate. But one thing is uncontested: cities have played a central role in the evolution of human civilization for thousands of years. Historically, the availability and types of resources in surrounding hinterlands limited the size and scope of a city. Water, fuel for energy, and building materials restricted a city's growth.

1

The physical geography—including vegetation, soils, landforms, topography, and climate—also shaped the social and economic characteristics of cities, determining how quickly cities grew, their patterns of expansion, building styles and construction materials, and even how much time urban dwellers spent outside.

Today, trade, commerce, and migration have decoupled cities from their immediate surroundings. Cities rely on both the local and global environments. Global supply chains bring resources from distant sources to urban centers. Migrant workers bring new labor and transform the cultural composition of even the most diverse cities. In the other direction, urban influence expands outward. Cities are not only regional hubs of production and innovation, but also global engines of growth, with economic and political effects on places and people far away. The environmental impacts of urban areas are also global. Urbanization has affected wildlife habitat and biodiversity, the global demand for energy, and greenhouse gas emissions and climate change. Cities today affect and are affected by the state of the entire planet.

Seeing the Forest and the Trees

New global impacts require new perspectives for understanding. Researchers have long studied urban regions by investigating from within the city. They have researched the individual person, building, street, block, or neighborhood. And while this close view is essential to understand local processes and interactions, it can obfuscate the larger environment in which cities operate and situate, and miss important patterns at regional or global scales.

It took from the beginning of human history to 1960 for the global urban population to reach one billion. It took another twenty-six years—until 1986—to reach two billion. Today, the global urban population is increasing by one billion roughly every thirteen years. One reason why urban populations are increasing at a faster rate than before is that increasingly, population growth is occurring mainly in urban areas. More than half of the global population lives in urban areas. In the United States and other industrialized countries, urban dwellers make up an even larger portion of the total population, often 80 percent or higher. To understand urban growth, however, we must look at both the population and the resources required to build, transform, and operate our cities. Every day, an area equal to 20,000 American football fields becomes part of the global urban landscape. Cities consume more than 75 percent of the world's primary energy. Cities and their built environments are literally transforming landscapes, reshaping landforms, vegetation, waterways, and regional microclimates. Our urbanizing planet is a story about human ingenuity, resolve, and perseverance. Given that we will add another two to three billion urban dwellers to our planet by the end of the century, it is critical that we have a broader perspective of cities.

Today, we have whole new means of seeing, studying, and understanding these complex connections between the urban environment and the world through data from satellites. Since the launch of Google Earth in 2001, satellite images of Earth have become easily accessible to the public. Many of us can see the most remote areas of the world with a single click, in stunning detail, even on the screens of our mobile devices. But it is not only the remote that fascinates us. Satellite data provide new perspectives on the pulse of humanity, and nowhere is this more evident than in urban settlements. This book takes advantage of these data to provide new perspectives on humanity's urban imprint and to tell the story of the city's rise.

Although the statistics about contemporary urbanization show unprecedented magnitude and pace of change, the view from space also shows us that urban areas take up a small fraction of our planet and that there is incredible diversity in an urbanizing humanity. We see that an urbanizing planet is composed of towns, villages, and other metropolitan regions of varying sizes and shapes. Urbanization today is not just about dense settlements and tall skyscrapers. Many urban dwellers around the world live in settlements that are relatively small in size—with several thousand inhabitants—challenging our preconceived notions of the city.

The images in this book situate urban areas in their larger environment. Our wider perspective shows cities as a critical driver of environmental, social, and economic change outside of city limits, and allows readers to consider cities as interconnected to their regional and global hinterlands. At the same time, we must never forget that cities must also be understood at the human level. We start each chapter more intimately, with short descriptions of our own journeys in urban developments.

Seeing in a New Light

When the United States government first launched nonmilitary environmental research satellites in 1972, only scientists interested in agriculture, forestry, and environmental monitoring had the access and knowledge to interpret satellite data. Even for researchers, data from the satellite Landsat used to cost over $4,000 per image. Now such images are freely available to anyone worldwide. With access comes responsibility—to make more informed decisions about how to design, plan, construct, and operate cities in better, healthier, more sustainable ways.

Satellites offer several perspectives impossible from unaided eyes. In some images in this book, we show repeat observations of the same place. These allow us to observe urban areas over time. What spatial patterns emerge as cities grow or shrink? How is the city changing in response to its environs and vice versa? How do changes in season and climate affect the urban landscape?

The images in this book also appear in unexpected colors. These alien landscapes are the result of a type of translation—a re-creation of light outside the visible spectrum in colors our eyes can see. Just as dogs can hear sounds inaudible to humans, sensors aboard satellites are able to "see" what our eyes cannot. Seeing in different frequencies can provide an advantage; butterflies use markings visible only in the ultraviolet (UV) spectrum to gauge the health of their potential mates; some birds use the UV reflectance in the skin of baby chicks to adjust feeding strategies; and for many fish, UV vision is essential for foraging and mate selection. Satellite sensors extend our vision into these frequencies and allow us to see the health of vegetation and the distribution of heat across a city, as well as urban attributes such as the types of materials used in buildings and roads. We also see the unexpected beauty of these urban centers and their surrounding landscapes— visible only from above.

Finally, the overhead perspective of satellite imagery offers a new context and thus new understandings. We see geometries of transport, diversity of forms, angularity and sinuosity of the urban settlements. We also see that, despite the contemporary trends and scale of urbanization, urban areas are relatively small globally but impactful.

We have organized the book into three sections that provide a narrative arc of an urbanizing planet. We begin with "Earth's Terrains," which showcases urban areas situated in their larger natural landscapes. In this section, we see cities defined by their physical environments such as mountains, rivers, and agrarian landscapes. In the second section, "Urban Imprints," we see the physical expression of urbanization. Cities and their supporting infrastructure such as roads,

urban design, and night lights leave tangible, material impressions on the planet. In the third section, "Transforming the Planet," we focus on how the demand for urban resources is changing landscapes, and how a dynamic earth is changing urban vulnerabilities.

The Future of the Urban Species

We can study an organism by looking at its component parts—we can learn about its individual organs, its heart or brain, or its systems, circulatory or nervous. But to understand it more completely, we must look at the organism as a whole—seeing how it reacts to its community and environment. By investigating cities as a whole, and comparing examples of cities across the globe, we hope to find, in some ways, the equivalent of an urban genetic code, an understanding of how details on the smallest scales can impact evolution on the largest ones. We hope to understand how a city's design will impact how it will grow, its ability to meet energy and resource needs, and the ways it may come into conflict with its habitat.

If we are an urban species, we must ask questions about our cities. Where and how do we choose to live? How do our choices today affect the planet future generations will inhabit? How will urban regions continue to evolve within the world? The answers matter. Urbanization is a permanent change to the surface of Earth. Urban sustainability will be a prerequisite for planetary sustainability, not only because that's where most people live, but because that's also where the demand for the planet's resources—water, food, energy—comes from. Cities are vulnerable to many things—climate change, extreme weather events, lack of social cohesion, economic disruption. But cities are also extremely productive places that are rich in ideas, innovation, and culture.

We hope the images provided in this book, of urban landscapes on every continent and of wildly different forms, offer a glimpse of some of the tools that we use to investigate cities. We hope they create a new vision to see a path forward. We have choices. And with the power to predict, we hope to improve our decisions.

Karen C. Seto and Meredith Reba
New Haven, Connecticut
2017

views from space

Satellites allow us to view Earth from a unique perspective. Not only can we gaze at Earth from a vantage high in the sky, we are also able to inspect the same area day after day and year after year—to observe and measure how a patch of land or stretch of water transforms over time. Yet, the most unique aspect of satellite imagery is its ability to reveal what the unaided eye cannot see: wavelengths of light beyond the visible spectrum. By looking at other wavelengths, we can explore the health of vegetation, the water content of soils, the heat emanating from a concrete surface, or the pollutants in the atmosphere. This new vision allows us to measure change, find patterns and disorder, track similarities and differences, and appreciate the unexpected beauty of familiar scenes.

Urban areas expand and take shape in expected and unexpected ways. We can see ordered city streets, growth spreading outward from urban edges, and development constrained by surrounding topography and other natural features such as waterways and coastlines. Satellite imagery allows us to see the beauty of cities, while also revealing some of the challenges and opportunities facing our urbanizing planet.

Bird's-Eye View
Orbiting Earth, sensors aboard satellites capture images of our planet's surface. Depending on the sensor, you may be able to see individual trees and cars lining the streets, or a city expanse may appear no wider than your finger. These images are—similar to a still life—a glimpse

of our dynamic planet. Sometimes we want to compare these snapshots of Earth's surface by viewing them on a computer screen or printing them on paper—in the same way that we use a map to navigate an unknown area of the world. But we need to know how big an area each snapshot represents on Earth's surface. Scale is usually recorded as one unit of measurement on the map, printed image, or screen. If we had a one-to-one (1:1) map of the world, one unit of measure on our map—let's say one centimeter—would also measure one centimeter on the ground. Basically, we could picture our map or image like a blanket covering Earth, or the peel of an orange (if the inside of the orange were Earth). Clearly, maps or images this large wouldn't be that useful. So, most of the time we assign a scale where one unit on the map or image equals many, many of those same units on Earth. This way, we are able to shrink down what actually exists so we can see more. The images in this book are captured at various scales—from the closeups of the city streets of Barcelona, Spain, to the urban footprint of Chandigarh, India, to the surrounding urban landscape of Shiraz, Iran. The colors in these images may seem unusual—they are chosen to highlight unique features of these cities and their surrounding environments.

Scale

Each of these snapshots captures a city at a different scale. The blues and greens of the Barcelona image display the street networks and clustered development of the Eixample district of the city. In Chandigarh, the city street network is again depicted in blue, but now we have zoomed out from the individual street level and can see a glimpse of the nearby surrounding topography. Zooming out even further around Shiraz, pictured in blue, we see the larger landscape of nearby mountain chains, water bodies, and regional vegetation.

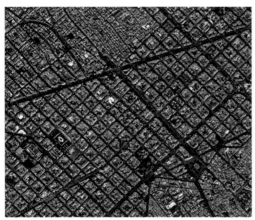

Barcelona, Spain
Worldview
1:40,000

Chandigarh, India
Landsat OLI/TIRS
1:496,000

Shiraz, Iran
Landsat OLI/TIRS
1:1,328,000

Shiraz, Iran

Location Map

Sometimes as we zoom out from the city center and it becomes smaller and smaller, you may not immediately know where to look to find the city. Locator maps, like the one pictured to the left, are a cue for the eye to spot the urban centers in the images.

Beyond the Naked Eye

We know that light travels as a wave. And that wave has a certain size, known as its wavelength. Shorter wavelengths (or higher frequencies) of light contain more energy.

The white light that shines from the sun consists of a range of wavelengths in the electromagnetic spectrum. Most of the sunlight that reaches the planet's surface is in the visible spectrum—wavelengths in a range from about 400 nanometers (which appears violet to our eyes) to 700 nanometers (which appears red). The colors of light that we can see represent a very small fraction of the entire spectrum. If our eyes could see different wavelengths, our world could look much different from what we see today—perhaps more like some of the images in this book.

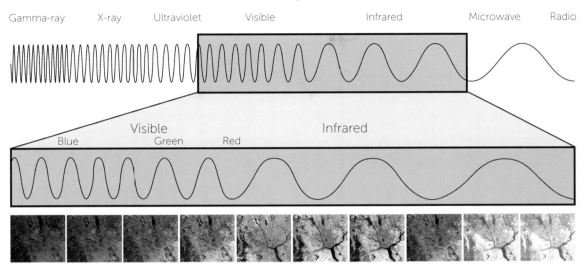

Electromagnetic Spectrum

Gamma-ray X-ray Ultraviolet Visible Infrared Microwave Radio

Visible Infrared

Blue Green Red

Satellite sensor data by wavelength

Electromagnetic Spectrum

Images of Yangon, Myanmar are spread across the bottom portion of the illustration. Each of these images, recorded by satellite sensors, shows what you might "see" at the exact same location, if you could only see at that corresponding wavelength. Notice how the same part of the image might reflect certain wavelengths of light very strongly (appearing bright white) but others not at all (appearing black).

True Color
(Visible red, visible green, visible blue)

(Visible green, visible blue, visible red)

(Near infrared, visible red, visible green)

Visible red

Visible green

Visible blue

Visible green

Visible blue

Visible red

Near infrared

Visible red

Visible green

Color Images

The satellite images in this book combine three different wavelengths of light to create color images. We can show these three wavelength ranges using three primary colors red, green, or blue, as indicated on the right of each image. The top, left image displays what our eyes might see if looking down from the satellite. We have printed red wavelengths of light with red ink, green wavelengths with green ink, and blue wavelengths with blue ink. In the middle, left image, we have printed green wavelengths of light with red ink, blue wavelengths with green ink, and red wavelengths with blue ink. Notice how clearly the city appears in green. In the third image, we go beyond the wavelengths that our eyes can see, printing the typically invisible infrared wavelengths with red ink. Vegetation reflects strongly in the infrared, and appears in bright red in this image.

Sunlight travels through Earth's atmosphere and is reflected off the planet's surface. Different materials on Earth's surface—grass, plants, concrete, wood, water, trees, metals, asphalt—reflect and emit light of different wavelengths back into the atmosphere and into space. Earth's atmosphere allows certain wavelengths of light to pass through it. These atmospheric windows are essentially transparent to the specific wavelengths of light. Remote sensing satellites and sensors often take advantage of these windows and the wavelengths of light passing through them by capturing the emitted or reflected light to record an image of Earth's surface. We can then display this image using colors that our eyes can see. For example, thermal infrared is a wavelength of light indicating heat and is not something we can normally see, but if we make this wavelength red in satellite imagery, the redder the section of an image, the higher its surface temperature.

By displaying these invisible wavelengths, we get a fresh view of familiar scenes. Urban areas become pink, turquoise, or blue. This technique helps us to emphasize different ways land has been used or developed within an image—such as land cover indicating vegetation, urban centers, or agricultural practices.

We use five main wavelength combinations in this book to illustrate urban areas. Example images of these wavelength combinations are shown on the following pages. As you become more familiar with these types of images, finding urban centers will become easier. We also include the names of the wavelengths to help you remember which wavelengths of light are displayed.

Wavelengths naturally visible to our eyes:
• visible red
• green
• blue
• panchromatic (used for sharpening, black and white)

Nonvisible wavelengths:
• near infrared
• shortwave infrared
• thermal infrared (surface temperature)

Berlin, Germany, in true color (visible red, green, blue).

near infrared, visible red, green

visible red, blue, near infrared

near infrared, shortwave infrared, panchromatic

shortwave infrared, near infrared, green

shortwave infrared, shortwave infrared, visible red

Wavelength Combinations
Berlin, Germany

Here, all of the five wavelength combinations used in the book are shown for the same city to help you to visualize the color palette of urban areas (center of the image), forested lands (northwest and southeast of city center), and agricultural areas (rectangular plots in the northeast of the image) for each different wavelength combination.

Notice how various combinations emphasize particular aspects of the city. For example, some wavelength combinations highlight differences between barren land and covered land, between urban areas and forested ones, and even between different types of vegetation.

Urban Structure

Below are images of Prague, Czech Republic displayed in four different wavelength combinations. The first image shows Prague in true color (visible red, green, and blue). The city appears gray. Vegetation appears in different shades of green, and barren agricultural fields appear in various shades of tan and brown. The second image shows Prague in the near infrared, visible red, and green wavelength combination often used to highlight vegetation such as croplands or forests. Vegetation appears in varying shades of red, helping us to differentiate types of vegetation, vegetation health, and growth stages. Fallow fields appear in shades of green. The city is shown in blue. The third image uses the shortwave infrared, shortwave infrared, visible red combination often used to highlight urban areas in purple. Vegetation is captured in different shades of green. Because this wavelength combination uses two shortwave infrared wavelengths, which are less affected by haze in our planet's atmosphere, it can look sharper than some other combinations. The

Wavelength Combinations

Prague, Czech Republic, is displayed in four different wavelength combinations.

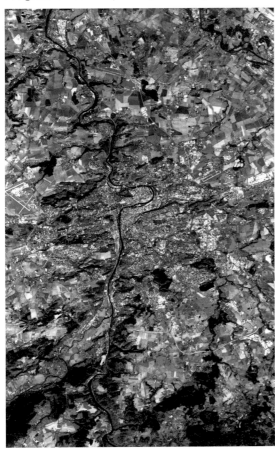

visible red, green, blue (true color)

near infrared, visible red, green

shorter wavelengths of light scatter easily in Earth's atmosphere, which is why we see the sky as blue. Finally, the last image uses the near infrared, shortwave infrared, panchromatic light combination. Here, it is more difficult to differentiate vegetation types, but the urban center is clearly pictured in blue. Because this combination uses the panchromatic wavelength, which has a sharper spatial resolution, the city structure can sometimes appear more detailed.

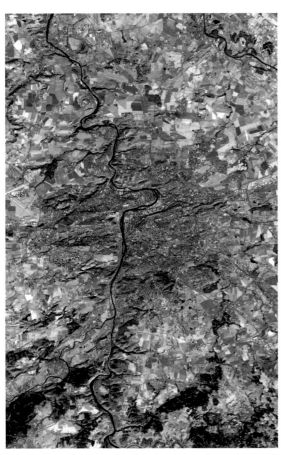

shortwave infrared, shortwave infrared, visible red

near infrared, shortwave infrared, panchromatic

visible red, green, blue

near infrared, visible red, green

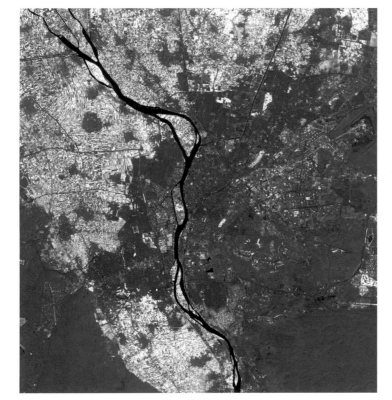

normalized difference vegetation index

Vegetation
Cairo, Egypt

Cairo is pictured in both true color and near infrared, visible red, and visible green in the two images above. Vegetation is clearly highlighted in the second image. For the image to the left, we calculated what is called the normalized difference vegetation index (NDVI) to emphasize vegetated areas. Regions with abundant vegetation are pictured in white, while areas with little vegetation are shown in gray and black.

Vegetation

Satellite images are often used to measure changes in vegetation—forests, agricultural lands, marshes. Being able to highlight the location of vegetation is also useful to help determine where urban areas are not. Here, we show Cairo, Egypt, in true color and again in near infrared, visible red, and green to spotlight vegetated areas. The different hues of red correspond to different types, health, and growth stages of vegetation in the image. These areas are easily distinguished from both desert areas and urban areas. The ways land is used in the image become clearer. If we want to define vegetated areas more drastically, we can calculate the normalized difference vegetation index (NDVI) for the image as well, which highlights the wavelengths of light essential for photosynthesis and living green vegetation. The image pictured at the bottom left of the opposite page highlights highly vegetated areas in white and nonvegetated areas in black. Taken together, all three images help us determine more precisely where vegetation is—and is not—in the area.

Here, we see three images of Paris, France, and its surrounding landscape. The first image below shows the city in natural color. The second image uses the light combination of near infrared, visible red, and green, which shows healthy vegetation in vibrant red. The urban center of Paris appears in blue. Barren agricultural fields appear in light green.

The last image shows heat, specifically surface temperature, over the exact same area. It uses only the thermal infrared wavelength of light. Here, we can see that highly vegetated areas (dark red) have a cooler surface temperature, while barren fields have a very high surface temperature, as does the airport near the

center of the image. The urban center of Paris has a temperature in between. Remote sensing allows us to visualize heat and surface temperature—something we cannot do with our eyes alone.

Paris, France, in true color.

Thermal Infrared
Paris, France

Here Paris is captured in both near infrared, visible red, and green and using only the thermal infrared wavelengths, which shows surface temperature (bottom).

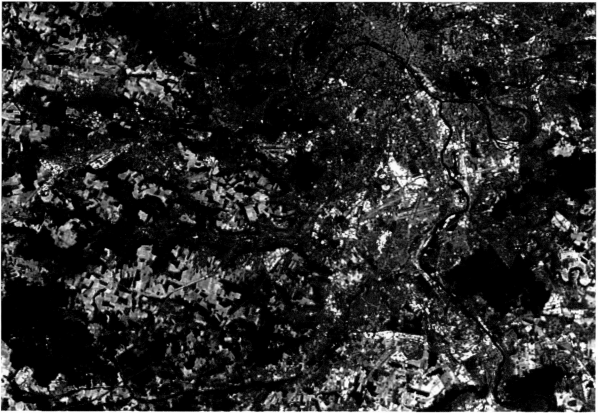

Spatial Resolution

Different satellites capture images at different magnifications. In remote sensing, we call these different levels spatial resolutions. An image's spatial resolution refers to the size of a detail of the pixel or cell in the image. For example, an image with a spatial resolution of thirty meters means that the pixel is measuring a thirty-meter by thirty-meter area on Earth. Objects on the ground that are smaller than thirty meters in size, such as a narrow alley or a car, would be difficult to see in such an image.

In this book, we use imagery from five main satellites:

- Landsat, provided by the National Aeronautics and Space Administration (NASA) and the United States Geological Survey (USGS), 30-meter spatial resolution
- ASTER (Advanced Spaceborne Thermal Emission and Reflection Radiometer), NASA satellite/Japan sensor, 15-meter spatial resolution
- CBERS China Brazil Earth Resources Satellite, 20-meter spatial resolution
- WorldView, DigitalGlobe (nongovernmental private company), 0.46-meter spatial resolution
- VIIRS (Visible Infrared Imaging Radiometer Suite), NASA, 750-meter spatial resolution

When we try to observe an image at a spatial resolution greater than it can resolve, images appear blurry or pixelated. Similar to images from a digital camera, the more pixels you have in a given area, the sharper the image.

To the right, we compare imagery of the Infosys campus, an information technology and business consulting company in Mysore, India. High-resolution WorldView images are shown on the left, and lower-resolution Landsat images on the right. The top series of images shows a zoomed-out area of the same exact region. Even at this scale, the differences in spatial resolution are clear. As we zoom into the area in the bottom row of images, the Landsat image becomes very blurry. We can actually see the square-shaped outline of the pixels the satellite sensor is able to resolve or distinguish. The detail of everything within these thirty-meter-sized squares is lost, as the square or pixel appears as one solid color.

WorldView, DigitalGlobe Image (true color)
0.46-meter spatial resolution

Landsat OLI/TIRS (true color)
30-meter spatial resolution

WorldView, DigitalGlobe Image (true color)

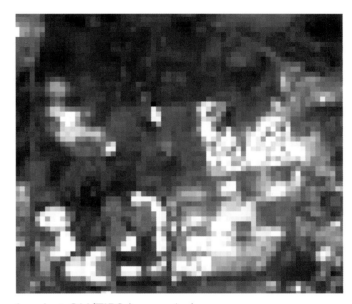

Landsat OLI/TIRS (true color)

Comparing Spatial Resolutions
Mysore, India

Here, we can see high-resolution images on the left, compared with lower-resolution images on the right. Can you spot how the buildings clearly spell out INFOSYS, the name of an information technology company?

You can observe a smaller difference in resolution in the images of Toronto, Ontario, Canada, below and on the next page. Here, we have two images covering the exact same area, one Landsat image with a spatial resolution of thirty meters, and an ASTER image with a spatial resolution of fifteen meters. Though at first glance the images appear very similar, careful inspection illuminates the more clearly defined street network in the ASTER image on the next page and better definition in the downtown urban core in the image center.

Looking at these examples, it may initially appear that higher spatial resolution, or a more detailed image, is always better. However, this is not the case. For example, if we are only trying to determine the general boundary of an urban area, but attempting to do this for many cities across the globe, a lower spatial resolution is actually preferable. Higher-resolution images are larger and require increased processing power and computation time.

Landsat OLI image: 30-meter resolution (near infrared, visible red, green)

Spatial Resolution
Toronto, Ontario, Canada

Toronto is shown here in turquoise,
while the surrounding vegetation
appears in red. The water of Lake
Ontario is captured in black.

ASTER image: 15-meter resolution (near infrared, visible red, green)

Time

A satellite orbits Earth, passing over the same place at a regular interval. Depending on the height, speed, and number of satellites collecting imagery, an image of the same place could be captured daily, weekly, or at some other regular interval. This allows us to measure change on our planet and within our cities. Depending on the time interval and the resolution of the satellite, this change could be urban expansion or decline, increases or decreases in road connectivity, spread or retreat of vegetation, or cooling or warming of surface temperature.

In the images to the right, we see Phoenix, Arizona, in July 1985 (top) and July 2016 (bottom). Here, the sprawling urban expansion over the three decades is clear. By comparing these images systematically, we can not only measure the area of growth, but also characterize the shapes and patterns of the city's spread. And when we compare these changes for cities around the world, important patterns at local, regional, or global scales sometimes emerge.

These repeated observations allow scientists to measure changes on Earth's surface from year to year, but also day to day or season to season. The same area can look quite different, for example, in the temperate latitudes from winter to spring—when deciduous trees have lost their leaves to when the new buds are blossoming. We see this clearly in the images on the following pages of Boston, Massachusetts. In the spring and summer images, healthy, abundant vegetation is pictured in red. The orange colors in the autumn image show dropping leaves, and snow is captured in the white and gray hues of the winter image. In all other seasons, the concrete and built-up urban surfaces appear turquoise. These seasonal changes remind us that when comparing a region or city over time, it is also important to consider the season or even the exact date the image was captured. We can also use remote sensing to compare the onset of different seasons from year to year. For example, we can determine when the leaves blossom on trees in the spring and when they fall in autumn. Comparing this data from year to year over an extended time period, we can see seasonal shifts and variations for particular geographic regions.

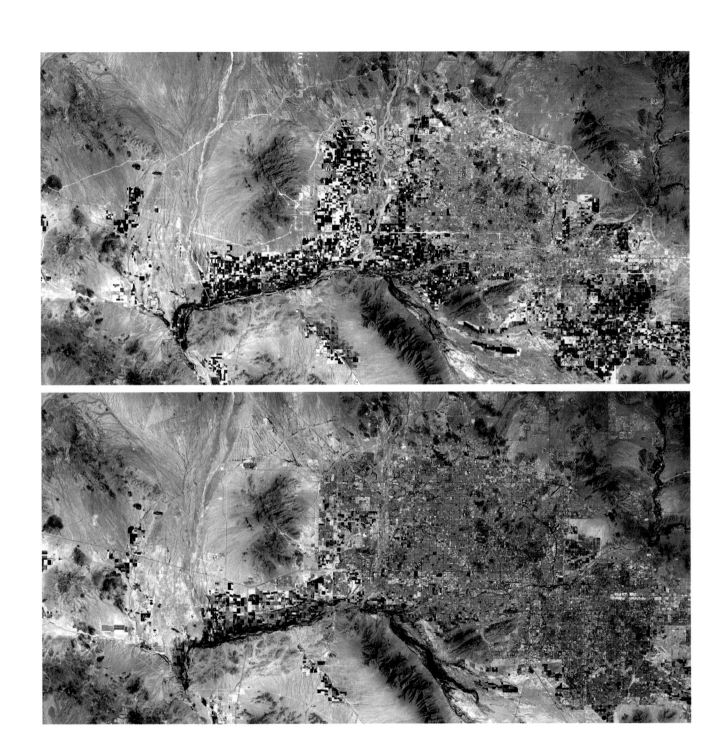

Urban Growth
Phoenix, Arizona, United States

The urban growth of Phoenix is evident in these two images taken thirty-one years apart (July 23, 1985, top; July 12, 2016, bottom).

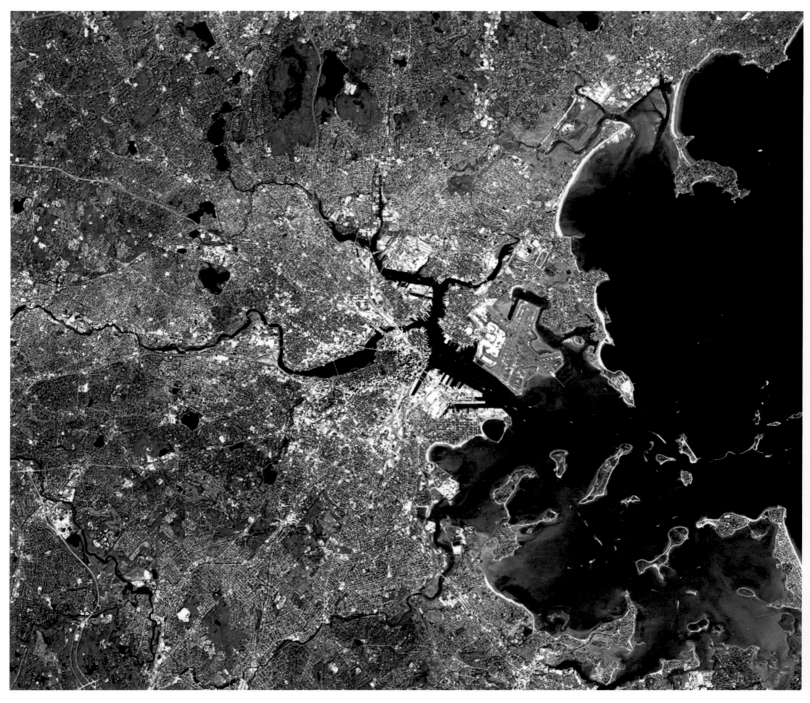

Boston, Massachusetts, United States in true color
(visible red, green, blue).

Winter: February 2, 2015

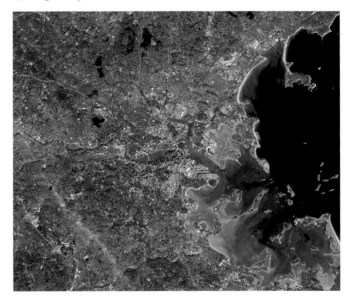

Spring: May 9, 2015

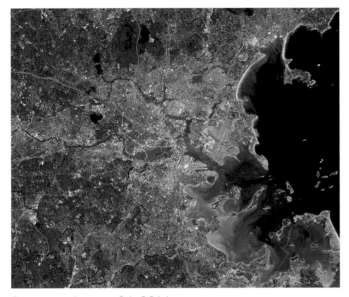

Summer: August 24, 2014

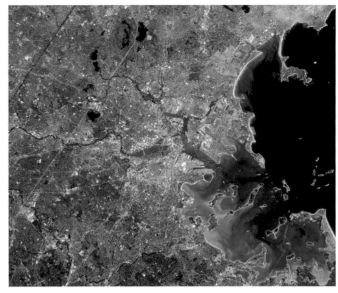

Autumn: October 31, 2015

Seasonality
Boston, Massachusetts

Boston is shown in each of the four seasons using the same channels of light (near infrared, red, green). Here, we can see that seasonal changes are also captured in satellite imagery. We also notice that the reflectivity of the city itself changes when covered by snow in the winter image. These changes are important to remember when comparing the same region over time.

On the following pages, we feature satellite images of urban areas from across the world, highlighting both the opportunities and challenges of our urbanizing planet. We have organized the book into three themes—Earth's Terrains, Urban Imprints, and Transforming the Planet—that reflect the arc of the urbanization process. We build cities in diverse terrains around the world, whether in mountains or near water bodies, creating symbiotic relationships with the natural environment. Agricultural landscapes, where the built and agrarian environments combine to create unique mosaics, are a necessary precondition to enable larger urban settlements. Urban areas leave a physical imprint on the planet through their built environments, laws, and land uses that create artificial boundaries, transportation networks, or urban design. No city stands still. Urban areas are constantly changing and transforming. Vast resources from around the world are required to build, operate, and maintain cities. Urban settlements have always been susceptible to natural disasters, but climate change, natural hazards, and other environmental changes are bringing the vulnerability of urban places into sharper relief.

Ultimately, the survival of the urban species will depend on how we can adapt to these challenges.

earth's
terrains

alpine

riverine

agrarian

When we view cities against the backdrop of nature's canvas, we see that humans have created urban settlements in the most unlikely of landscapes: rugged, treacherous places that are hostile, barren, and cold; isolated territories that are dusty, sweltering, or full of thick, sticky air. Sometimes these cities' placements seem illogical—they sit precariously in places vulnerable to natural hazards of fires, earthquakes, landslides, and tsunamis.

At their best, cities do not subvert their surroundings but embrace them. Rivers, valleys, mountains, and deserts all influence the form and feel of a city. Settlers in rugged mountainous landscapes sought spiritual inspiration from their majestic views. Some towns grew into cities because of the awe that the physical landscape inspired. Seeing cities in their larger landscape, we observe and appreciate cities differently.

A number of physical features shape the city: presence of mountains, rivers, quality of soil, and presence and distribution of plants and animal life. In turn, these different features affect the suitability and habitability of different landscapes. Over a short period of human history, the urban species has spread over the entire planet. Yet, despite the magnitude of the trends of contemporary urbanization—three billion urban dwellers will increase to seven billion by the end of the century; with a city of one million being developed every ten days—we see that urban areas occupy only a small fraction of the planet and that they cannot escape the natural conditions in which they find themselves.

In this section, we present images of urban areas in some of the most undeveloped and rugged terrains around the world. We have organized the section across a range of landscapes, starting with montane environments, where the scale of the physical environment always exceeds the magnitude of urban development, continuing with river systems, where towns, cities, and villages must learn to live with the undulating pulse of the water, and concluding with agricultural landscapes, which are the result of tightly coupled interactions between human-modified systems and their natural environment.

alpine

alpine

It was a motley crew, made up of nearly fifty people—climbers, guides, porters, cooks, mule drivers, and trekkers. At the moraine camp, we made a makeshift stretcher, while another group climbed to the high camp at an elevation of about 16,000 feet to assess the patient.

Stabilizing the injured climber on the stretcher at the moraine camp was only the first step. Now began the hard part: a thirty-hour hike out to the nearest road. We clambered down the loose talus and scree, trying to avoid a slip and unbalancing the stretcher. It had been hours since we ate anything substantial, but no one said a word. Day had long since turned to night.

Suddenly, off in the distance we saw a light. The local mule drivers, porters, and cooks, who earned their living cooking and hauling gear for climbers along the two-day trek to base camp, led us off the trail toward a traditional tent-like structure their families had set up in the middle of the river valley to feed us all. And today what remains most vividly in my mind from these days is not the actual rescue of the climber, but this alpine community's ingenuity, their strength, their kindness and willingness to help make life easier for us all.

French Direct Route, Alpamayo, Cordillera Blanca, Peru
July 2010

Mountains challenge us mentally and physically. They test us and make us resilient. Mountains cover about one-quarter of Earth's land surface and are home to about 12 percent of the global population. Mountain communities are among the world's poorest and most vulnerable. Yet, they are also home to some of the world's most resourceful and resilient groups, since they are often located far from city centers with limited access to infrastructure in harsh climates. Around the world, alpine communities are undergoing two disparate trends: rapid growth or rapid decline. Labor migration, agricultural abandonment, and growth of religious, ecological, and adventure tourism have created or supported the growth of some urban settlements, while others are being abandoned.

In some of the most fragile alpine environments, the growth of urban areas, which leads to the construction of roads and dams, could increase the vulnerability of mountain communities to extreme climate events and other environmental hazards. At the same time, the development of transport and energy infrastructure is both connecting and transforming these once remote communities. Building construction, deep road cuts in steep hillsides, and unplanned urbanization, all of which require cutting into bedrock or crossing geologically weak areas, have resulted in increased and more severe occurrences of hillside collapse, landslides, debris flows, and rock slides that put millions of people at risk. A 7.8-magnitude earthquake in Nepal in 2015 killed more than 9,000 people and injured more than 23,000. Floods in Uttarakhand, India, in 2013 left nearly 6,000 dead and more than 100,000 people trapped by landslides, damaged roads, and flooded conditions.

Ultimately, even human ingenuity is limited in how much it can move mountains. In the end, topography is the ultimate constraint on the patterns of urban development. These images show that cities and towns exist in some of the steepest, most rugged terrains. These images also challenge our notion of urban. Base camp in Mount Everest doesn't have a city hall, but it is central in providing life-support services to the hundreds of people who attempt the climb and the more than 40,000 who reach the camp.

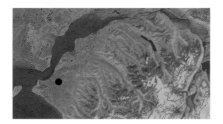

Anchorage, Alaska, United States (next page)
Landsat OLI/TIRS
1:415,000

The Chugach Mountains fill the right half of the image with the two arms of the Cook Inlet—the Knilk Arm (north) and Turnagain Arm (south) in the left side of the image—surrounding the city of Anchorage (pictured in orange) perched on the mountainous peninsula. The neon blue highlights the reflective snow on the top of the mountain range. Because both bare rock and city building materials, such as cement, often reflect light at a similar wavelength, both the rocky mountain outcrops and city center appear bright orange. Due to the city's location at 61.2° north latitude, the longest day of the year—the summer solstice—spans about 19.5 hours between official sunrise and sunset, as opposed to only about 5.5 hours on the winter solstice.

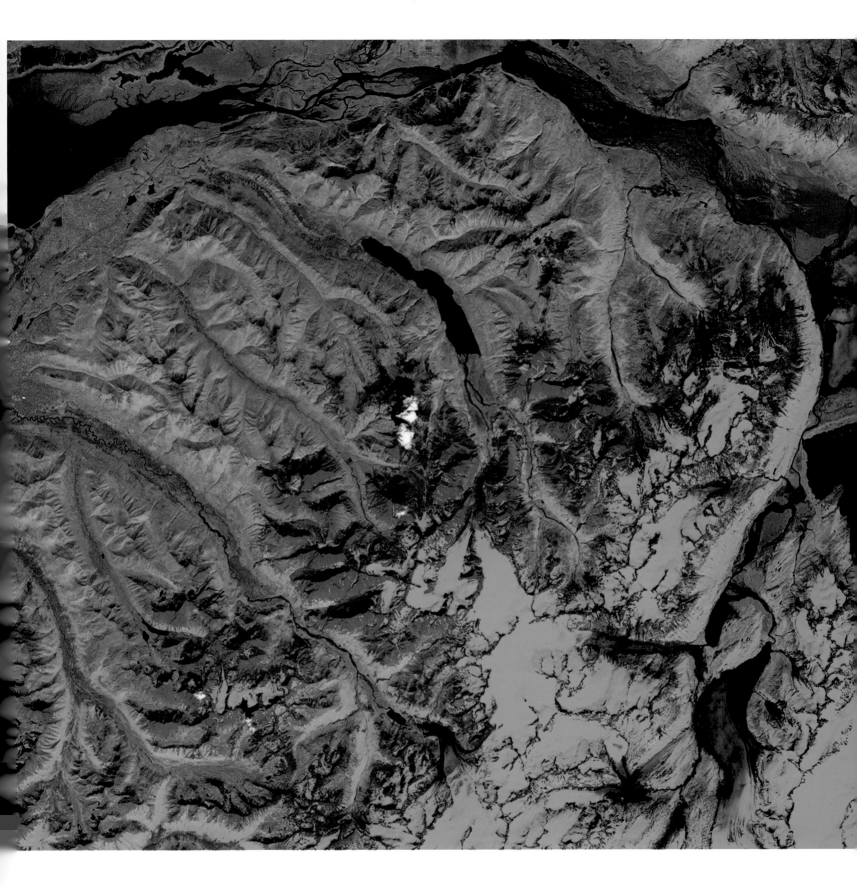

Gilgit, Pakistan
Landsat OLI/TIRS
1:173,000

Gilgit is nestled between three major mountain ranges—the Himalaya, Karakoram, and Hindukush. Surrounded by snow- and cloud-covered peaks at the major confluence of the Hunza and Gilgit Rivers, the city sits just off the Karakoram Highway and is interspersed with the red riparian vegetation to the west of the fork in the rivers. Clouds are visible forming above the surrounding mountain peaks. The region is generally dry, and melting glaciers supply water for agricultural irrigation. The surrounding steep mountain slopes make the region prone to landslides.

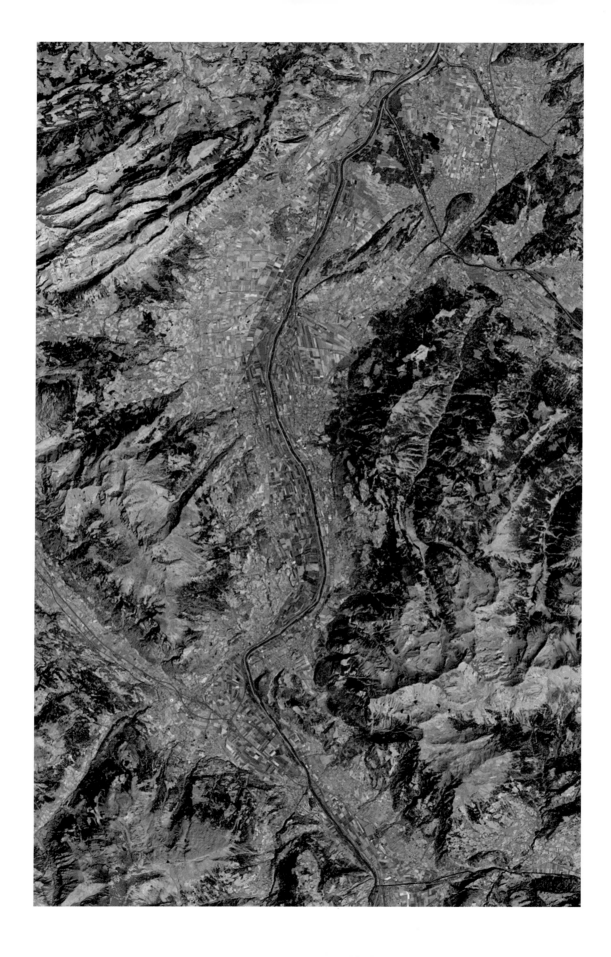

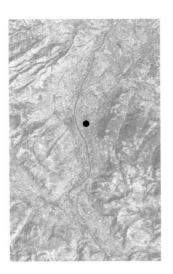

Vaduz, Liechtenstein
Landsat OLI/TIRS
1:178,000

Situated on the banks of the Rhine River, Vaduz
is dwarfed by two mountain peaks to the east,
Gafleispitz and Alpspitz. Consistent southerly foehns,
dry down-slope winds, bring precipitation and
warmth to the lee (downwind) side of the mountain.
This allows otherwise unlikely crops such as corn
and grapes to grow in this region. The city itself is
quite small, pictured in green on the eastern side
of the river in the middle of the image. The crops
appear as rectangular plots of oranges, blues, and
yellows. The different colors of crops correspond
to different health and growth stages.

Shiraz, Iran
Landsat OLI/TIRS
1:418,000

Located in a fertile valley that was historically home
to many vineyards, Shiraz was an important medieval
Islamic city during the Zand dynasty (1747 BCE–79 CE)
and the capital of Iran from 1750 to 1794 CE. The city
is located at the base of the Zagros Mountains, where
a seasonal river just north of the city flows into Maharlu
Lake in the center of the image. The rugged semiarid
landscape surrounding the city appears as a grooved
and textured landscape in the image.

Bamako, Mali
Landsat OLI/TIRS
1:250,000

Bisected by the Niger River, the generally dry and dusty
city of Bamako (pictured in a purple-blue) sits at the
foothills of the Mandingo Plateau cliff, while the sparse
and concentrated riparian vegetation (depicted in red)
creates a weblike matrix throughout the landscape.
The city name itself, meaning "crocodile river" in
Bambra, the local language, reveals the importance
of the river.

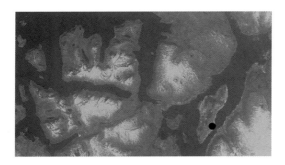

Tromsø, Norway
ASTER
1:206,000

Norway's capital of the north, Tromsø (pictured in an aqua blue) is spread over two islands, Troms (Tromsøy) and Kval (Kvaløy), and the mainland to the east. Most of the city is located on Tromsøy, in the bottom right quadrant of the image, and is surrounded by snow-covered peaks. The city is located about 250 miles (about 400 kilometers) north of the Arctic Circle, where the sun shines continuously from May 20 until July 20. The black regions depict the fjords that define the coast in this northerly latitude.

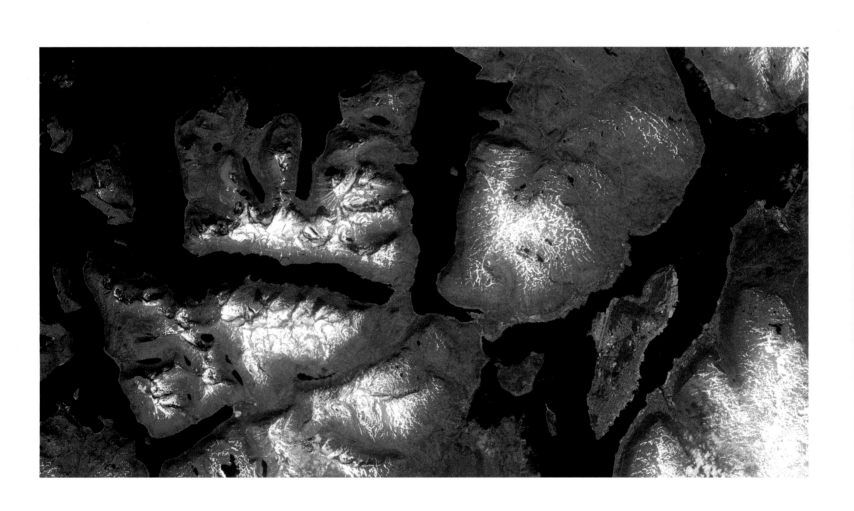

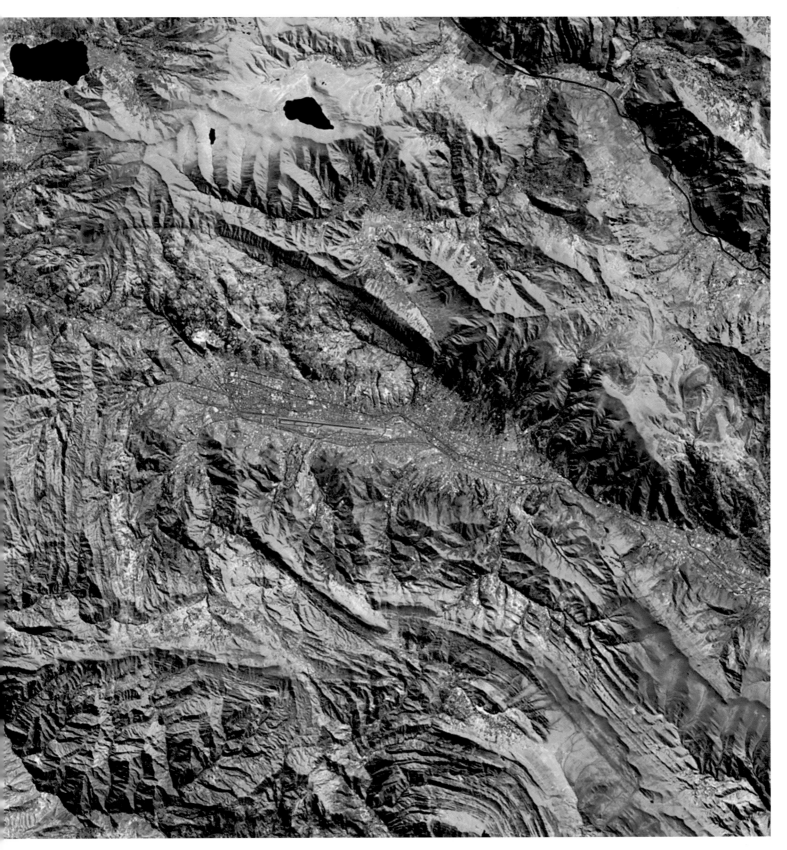

Cusco, Peru
Landsat OLI/TIRS
1:174,000

Situated at an average elevation of about 11,000
feet (3,400 meters) above sea level in the Huatanay
Valley of the Andes Mountains, Cusco (pictured
in blue) was an ancient Inca Empire capital. The
Willkapampa Mountains lie to the north. Due to
Cusco's mountainous location, roads throughout the
region are not well developed, and many tourists must
fly to visit this city. Much of the spatial organization
and buildings from the Inca Empire remain today in
the city footprint itself. One example is the region of
Saksaywaman, located in the white, slightly circular
region on the north/northwestern edge of the city in
the foothills. Saksaywaman is a historic Inca Empire
capital and is often visited by tourists who wish to see
its large, terraced stone walls.

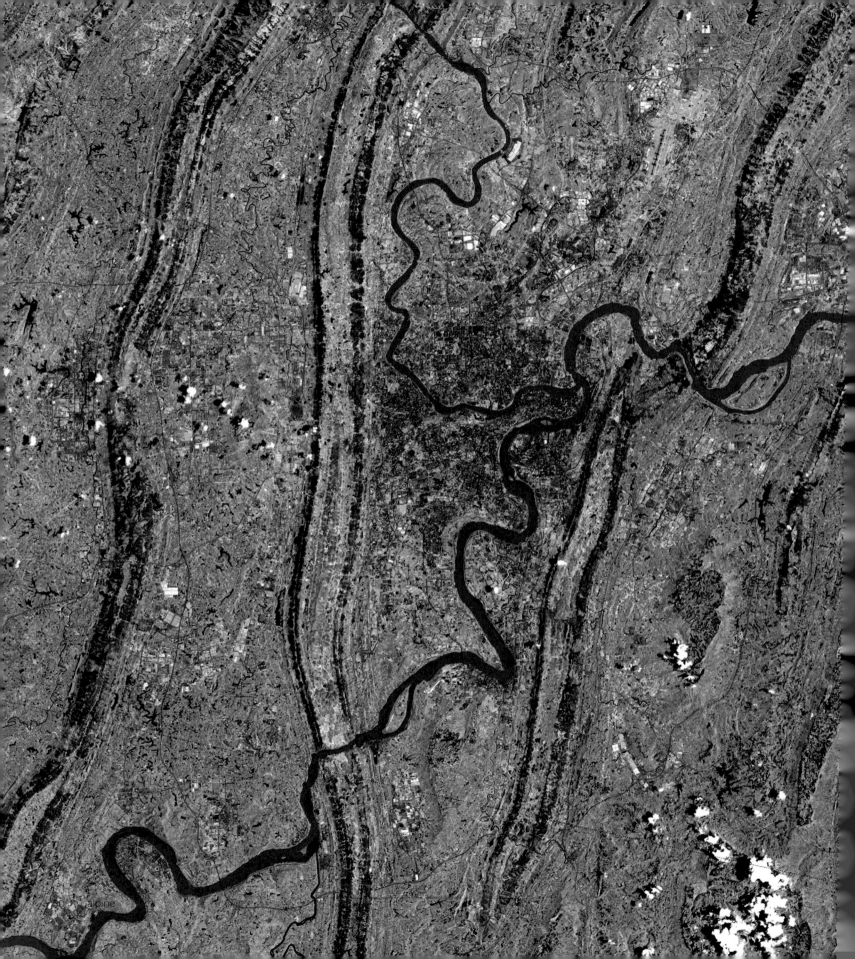

Chongqing, China
Landsat OLI/TIRS
1:277,000

Born in the ancient Ba kingdom, Chongqing originated
at the confluence of the Yangtze and Jialing Rivers.
Geologically, the city is built on red sandstone and
shale in a major coal-mining region in China. The
three main vertically oriented "grooves" in the image,
called anticlines and synclines, are folds in the earth's
surface rocks.

Osh, Kyrgyzstan
Landsat OLI/TIRS
1:175,000

Historically, Osh (pictured in purple) was a key stop along the trade route between China and India. Today, agricultural communities growing crops of tobacco, grains, cotton, and melon occupy the Alay foothills to the north of the city. The varying colors of the rectangular fields highlight the variety of crops and growth stages, while fallow fields are pictured in purple. The Alay Mountains border and constrain the city expansion to the south.

Niamey, Niger
Landsat OLI/TIRS
1:180,000

Niamey, the largest and capital city of Niger (purple), sits 700 feet (200 meters) atop two sparsely vegetated low-lying plateaus on the northeastern banks of the Niger River. The Kennedy Bridge in the center of the image was built in 1970 and allowed the city to expand to the south. Water carrying suspended particles and sediment appears in br ght blue in the top right quadrant of the image, while the edges of ridge-like plateaus (dark burgundy) form almost a dendritic pattern. The color of these rocky ridges and the city itself are nearly identical because certain rocks or bare soil reflect light in a similar way as city building materials such as concrete.

riverine

riverine

At around 1:00 pm, I joined the villagers working to dig ditches for water pipes throughout the settlement. The village was situated just off the Yarlung Zangbo River in an otherwise dry, high alpine expanse. Day after day, the villagers (usually women) walked to the river with an oblong plastic container rigged with makeshift woven straps. Here, they would dip the container in the river and strap it to their back for the return trek home. This was their water for everything—for drinking, cooking, tea, cleaning their homes and themselves. After years of making this trip, the village was now able to divert upstream water to run directly through the settlement, hopefully shaving hours off their daily routine.

Kampa La, Tibet
November 28, 2006

Life requires water. And when a settlement lacks reliable water access, nearly every other aspect of life is affected. Many of the oldest cities in the world have their origins in riverine landscapes: on rivers, stream branches, mouths of bays, harbors, along shores. Some cities are barely higher than sea level. Some are located at the confluence of rivers and tributaries; others sit in distributaries, where rivers branch off into smaller streams and create a delta. Local water availability limited the size of towns and cities. Living along a river meant living with seasonal change. Water systems shape riverine cities. Historically, towns and cities situated along rivers faced episodes of diseases or other health epidemics when sewage contaminated the rivers. Those living downstream of other cities were subject to water-borne illnesses.

Rivers can play a uniting or dividing role within a city. While some cities built levees and dykes for protection from the uncertainties of the water, others created a dependency on the water and cultivated agrarian livelihoods. Sometimes a river separates two different types of development, old and new. One side develops constrained by the river, while the other side seems to thrive in spite of it. The river can separate cultures, histories, and urban form or act as a border or system of defense. The river can also form a network between cities with goods and people flowing along its path. Today, 90 percent of the world's trade is still transported through waterways. Cars, cell phones, books, diapers, toothbrushes, and plates are loaded and unloaded at the world's ports.

These images show the symbiotic and sometimes contentious relationships between urban areas and their surrounding water systems.

Guayaquil, Ecuador
Landsat OLI/TIRS
1:250,000

Located in the largest estuary ecosystem on the Pacific Coast of South America, Guayaquil (shown in light blue) sits at the confluence of the Babahayo, Daule, and Guayas Rivers (shown in a blue-violet). The majority of Ecuador's trade passes on ships through this port. The highly vegetated marsh and agricultural land (red) borders the city to the southwest. Flooded fields appear black in the bottom half of the image because they have a very low reflectance and absorb most of the sun's light or radiation.

riverine

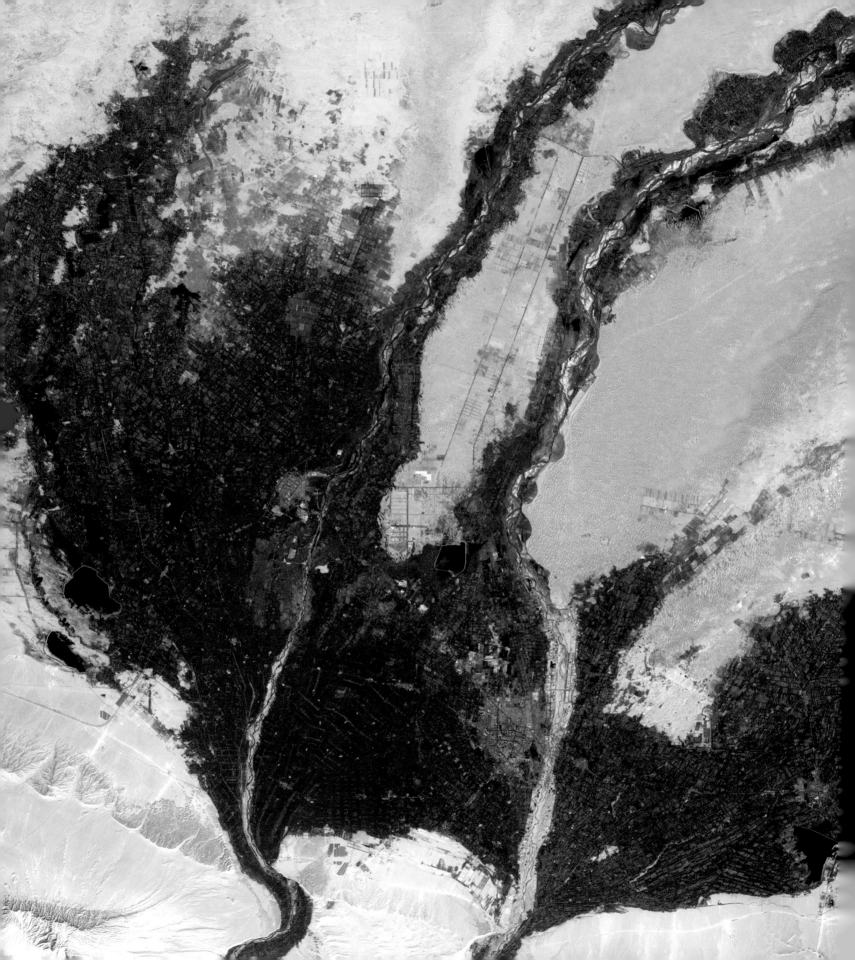

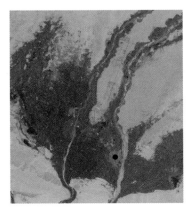

Hotan, China
Landsat OLI/TIRS
1:288,000

It's surprising to see this much vegetation in the middle of a desert. The city of Hotan (pictured in tan on the west bank of the easternmost river) is a center of irrigated cultivation and a historic junction point of the southern Silk Road trading route. Located in the Taklamakan Desert, the irrigated agricultural plots and heavily vegetated riparian areas surrounding the river (shown in red) create an oasis effect, helping to cool the city itself.

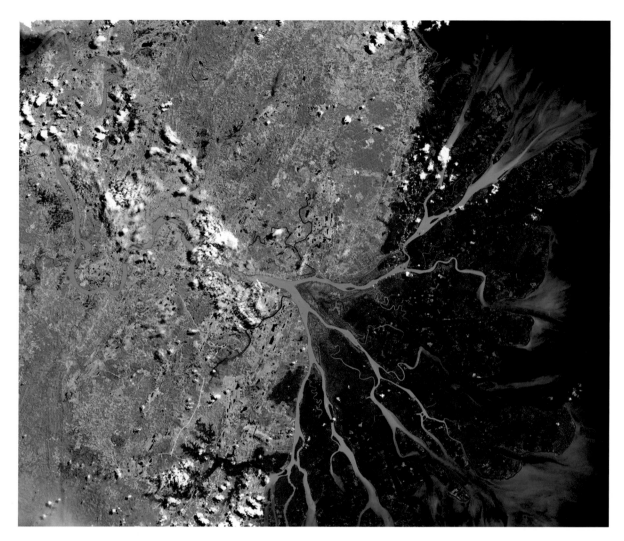

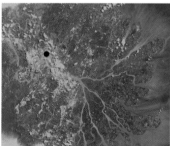

Samarinda, Indonesia
Landsat OLI/TIRS
1:510,000

The city of Samarinda thrives on the annual flooding of the Mahakam Delta. Lowland forest (pictured in orange) and shrimp ponds (shown in reddish-purple) surround Samarinda. It wasn't always this way—the shrimp ponds are replacing historic mangrove forests, and the border between the orange and red is bound to shift. The bright pink splotches along the top and bottom right edges of the image are clouds.

Manaus, Brazil
Landsat OLI/TIRS
1:330,000

Manaus is Amazon's largest city and one of its busiest ports. Despite its importance, only one cable bridge spans the river to the southwest of the city. The Rio Negro to its south connects Manaus to the world. The Rio Solimões appears in light turquoise because it carries more sediment, and sand particles suspended in its waters reflect light differently. The Rio Solimões flows north into the Amazon River in the far east of the image. The rectangular red patch to the northeast of the city marks the boundaries of the Adolpho Ducke Forest Reserve created in 1963 to honor the entomologist and botanist who was an expert on flora in the region.

Bandar Seri Begawan, Brunei
Landsat OLI/TIRS
1:217, 000

Bandar Seri Begawan, an agricultural center and port city (shown in light blue), sits along the northwest bank of the Brunei River. Brunei Bay (in blue in the center of the image) radiates into regions of both Brunei and Malaysia along the delta, including the Labu Forest Reserve of Brunei in the bottom center (in burnt orange), where a bright green line indicates a clear-cut path through the reserve.

Rosario, Argentina
Landsat OLI/TIRS
1:174,000

To the west of the Paraná River is the city of Rosario, a major port city of Argentina. To the river's east is a sprawling marsh. Bright red areas show both gridded agricultural lands (to the west) and threads of vegetation in the marsh (to the east). An even tinier thread, Route 174, emerges perpendicularly from the boundary—a lone road crossing the marsh and heading toward the smaller city of Victoria beyond the picture's frame.

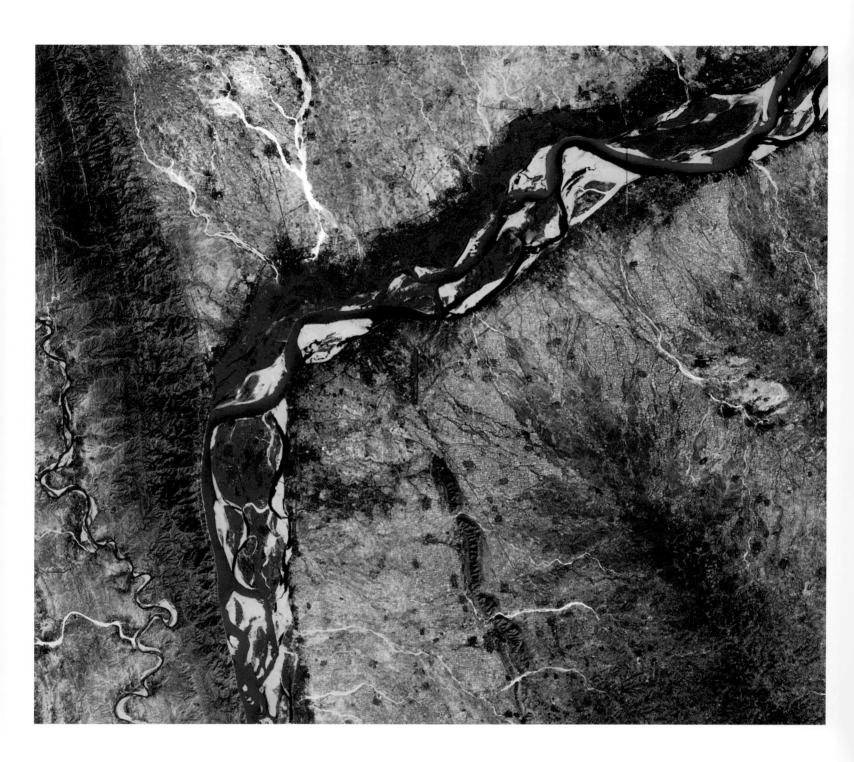

Bagan, Myanmar (Burma)
Landsat OLI/TIRS
1:244,000

Originally home to over ten thousand Buddhist temples at the peak of the Kingdom of Pagan from the eleventh to the thirteenth centuries, Bagan lies to the east of the Ayeyanwady River in the Mandalay region of Myanmar. Red highlights the highly vegetated land surrounding the river as well as small vegetated areas, which faintly dot the landscape.

Seoul, Republic of Korea
Landsat OLI/TIRS
1:148,000

Twenty-seven bridges span the Han River in the center of Seoul (pictured in blue), sewing and pulling the two halves of the city together and acting as networks between communities on opposite sides of the river.

The river itself used to connect Korea to China and the world via the Yellow Sea, but it is no longer used for such transportation since it crosses into North Korea.

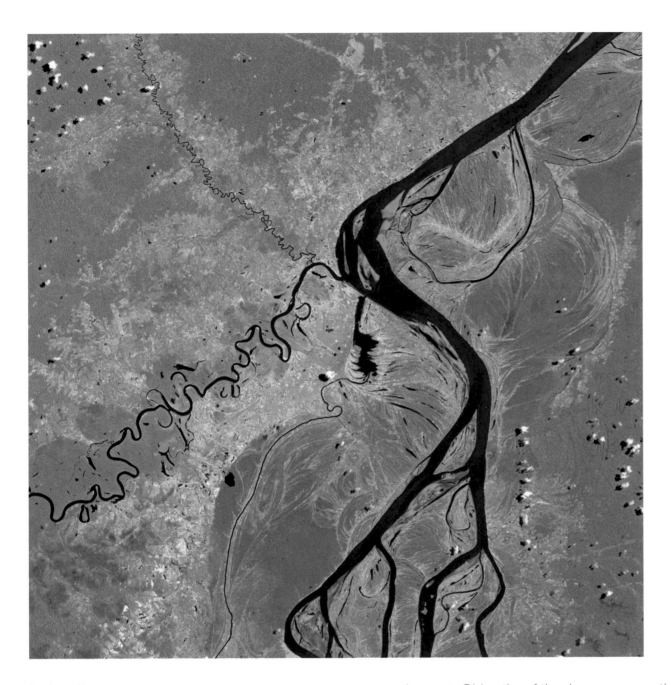

Iquitos, Peru
Landsat OLI/TIRS
1:260,000

Iquitos is the largest city in the world that cannot be reached by road. If you want to visit Iquitos (shown in light blue), you will have to come by plane or one of two rivers, the Amazon to the east or the Nanay to the west. Old paths of the river appear on this image in lighter orange, a vestige of where the water once flowed and a sign that rivers change their paths.

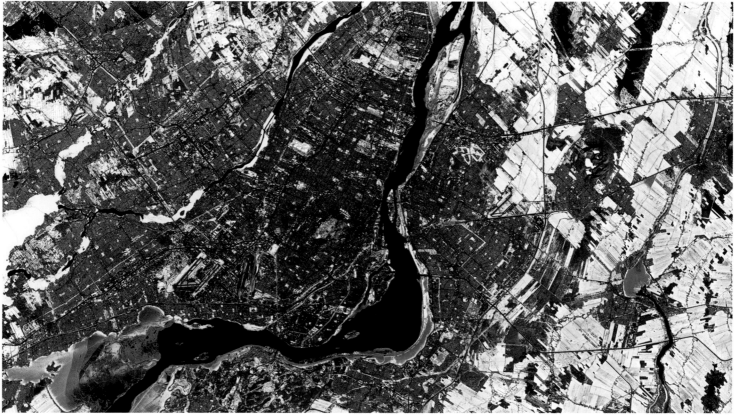

Montreal, Quebec, Canada (Summer)
Landsat OLI/TIRS
1:315,000

Four rivers flow near the city of Montreal in this image from east to west—the Rivière des Mille Îles, the Prairies River, the St. Lawrence River (largest), and the Richelieu River. The main portion of the city is on the Island of Montreal in the center of the image. After Montreal was incorporated as a city in 1832, the Lachine Canal, which can be seen on the eastern edge of the Island of Montreal, opened and allowed ships to bypass difficult rapids in the main waterway.

Montreal, Quebec, Canada (Winter)
Landsat OLI/TIRS
1:315,000

Here, Montreal is captured in winter. Although both images show the same frequencies of light, snow and ice are highly reflective and change the color of the picture. Ice is visible along the edges of the St. Lawrence River, and snow covers the surrounding narrow agricultural plots, called long lots. Because these fields are not vegetated in winter, they do not have the bright red color of the summer image. This image cannot capture the well-known tourist attraction called the "Underground City" or RESO, a collection of walkways connecting hotels, restaurants, shopping centers, and subway stops that covers an area of about 4.6 square miles (over 12 square kilometers) below the snow-covered city itself.

Khartoum, Sudan
Landsat OLI/TIRS
1:295,000

Bordered by the White and Blue Nile Rivers on its left and right, respectively, Khartoum has a roughly triangular shape. Highly vegetated riparian regions (pictured in orange) flank the rivers as rectangular agricultural plots, while a small region of circular "pivot" irrigated agricultural plots dot the landscape to the southeast. If the rivers form a boundary, so does the surrounding desert (shown in green).

agrarian

Over a dozen farmers wearing conical hats were walking back and forth along the coastal plot, about the size of a garden bed. They used flat rakes to push a white crystalline substance to the surface. They had moved the water along small narrow ditches to here from the sea about one week earlier, and the water was quickly evaporating under the hot sun, exposing the white residue. In a few days, they would collect enough salt to take to the local market. Although his family had been salt farmers for three generations, their buyers were changing. Now there was talk of overseas customers, restaurateurs, and city folk who wanted the salt for their traditional, small-batch, hand-harvested methods.

About 1,700 kilometers south of the salt farm, saltwater infiltrated the agricultural lands, ruining the rice crops. The water had become saltier and saltier each year and the farmer noticed the decline in yields, but this was the first year that the crops were completely lost. The soil on the fields was covered with a thin white layer of salt. Several of his neighbors had converted their rice fields to aquaculture, mainly shrimp. Others had stopped farming their land and took jobs in the nearby town. His family had been rice farmers for three generations, but he, too, was contemplating a city job.

Nam Định and Sóc Trăng, Vietnam
November 2003

Agricultural villages, where farming and agrarian activities dominate life, are some of the earliest types of human settlements and have existed for over ten thousand years. The ability to grow food allowed societies to flourish and grow. The itinerant and nomadic hunter-gatherer lifestyle was no longer mandatory for survival. Access to water limited what could be grown, how much, and when, while resource constraints like soil conditions, topography, and sunlight limited the size and location of agricultural plots. Human settlements could only grow as large as access to food and water would support.

For much of history, agrarian communities existed within a mosaic of agricultural plots. Small farm plot sizes reflected the capacity of manual labor, the limits of mechanization, and household size and structure. Cities were dependent on local or regional food production. This gave rise to regionally specific crops and diets. After center-pivot irrigation—a sprinkler system that rotates around a pivot point and waters crops in a circular pattern—was invented in the 1940s, it profoundly transformed agrarian societies. Today, we have agrarian societies that specialize in a single crop. These monocrop societies grow food for faraway places, and very little of the food is consumed locally. Increasingly, cities are more dependent on global food production. Local varieties and regionally specific food sources are giving way to global production. Agrarian communities need to redefine themselves to be competitive in global and regional markets.

And while some places now grow only one crop, other former farmlands grow none at all. The urbanization of farmland—conversion of growing land to residential and commercial landscapes—means a loss of not only growing potential, but also knowledge, livelihoods, and lifestyles. Some regions, such as China and sub-Saharan Africa, are experiencing a sharp loss of arable land. However, it is important to remember that places where urbanization peaked in earlier periods, such as Europe in the nineteenth century and South America in the twentieth century, also experienced a corresponding farmland loss and conversion during these earlier periods.

Semikarakorsk, Russia
Landsat OLI/TIRS
1:192,000

Semikarakorsk, shown in pink near the center of the image, is located in the Rostov region of Russia. Rostov is one of the largest agricultural regions in the country, with highly productive, thick soils. The patchwork shown in the image reflects the variety of grain crops. The river brings nutrients to these fields, but it also brings ruin—floods have forced the city to relocate at least four times during its history.

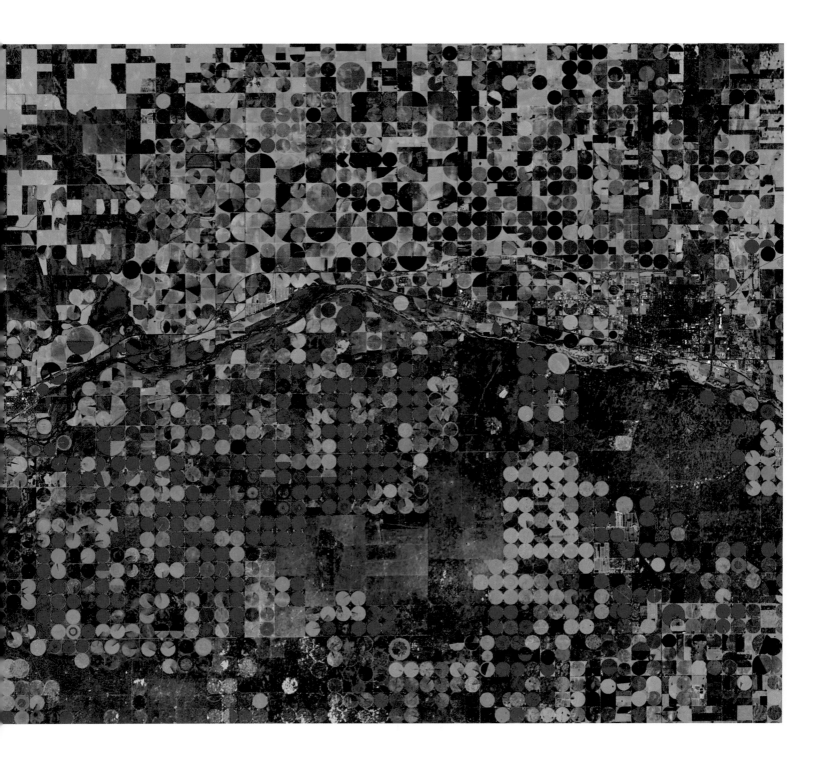

Garden City, Kansas, United States
Landsat OLI/TIRS
1:202,000

Frank Zybach, a farmer from Colorado, United States, invented "center-pivot" or circular irrigation in the 1940s to provide efficient irrigation of croplands. It became popular because it uses less water and costs less than traditional agricultural irrigation methods. The Arkansas River, which flows to the south of Garden City (pictured in purple in the right-hand side of the image), irrigates the main crops of corn, wheat, and sorghum. The satellite image captures different growth stages, crop types, and vegetation health. Less dense, newly planted crops, for example, appear in brownish green north of the river.

Bologna, Italy
Landsat OLI/TIRS
1:150,000

The well-preserved historic center of Bologna,
constructed mainly of red brick, is visible in the center
of the image (light blue color) and is surrounded
by the symmetrical spider-shape of the full city,
with its geometry the result of road transportation
infrastructure. Bologna is contained by the natural
boundary of the Apennine Mountains to the south and
a human-made boundary to the north, driving route
E45. City tourists often escape to the agricultural plains
to the north.

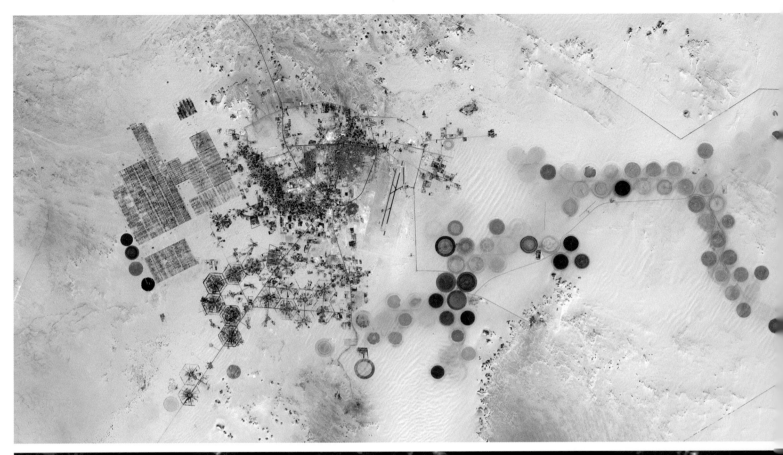

Al-Jawf, Libya
ASTER
1:205,000

Home of the Al Khufrah Oasis, this arid region is fed
by an underground aquifer. The city itself (gray) is
located in the center of the image and is surrounded
by vegetation pictured in red and green. Fallow fields
also appear gray in the image. Three different irrigation
methods are captured: center-pivot irrigation (circular
plots), a grid system in the western portion of the
image, and honeycomb-shaped plots just south of the
city—all attempting to maximize water efficiency in
this region of the world, which records only 0.1 inches
(0.254 centimeters) of rain annually.

Al-Jawf, Libya
ASTER
1:205,000

This image depicts the same area revealing surface
temperature—active crops are pictured in black
(where plant evapotranspiration cools the surface
temperature), while bare soil fields are pictured in
white/yellow as they absorb the sun's heat and lack
the cooling effects of evapotranspiration. The sandy
areas in the southeast appear dark due to their highly
reflective surface or albedo. The center of the city itself
has a variable surface temperature—with black regions
corresponding to cooler surface temperatures, most
likely highly vegetated regions.

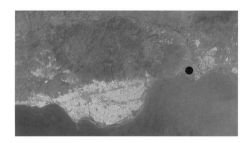

Almería, Spain
Landsat OLI/TIRS
1:350,000

The city of Almería (on the left side of the eastern, pointed peninsula) is barely visible in this image—but vast arrays of greenhouses covered in plastic (spanning more than 130 square miles, approximately 209 kilometers) appear in bright white. An otherwise constricted growing environment, Almería is Spain's strongest fruit- and vegetable-producing region.

Sapporo, Japan
Landsat OLI/TIRS
1:357,000

Sapporo (purple), flanked by Mount Teine and Mount Moiwa to the southwest, is a major agricultural region in Japan and is home to a wide variety of crop production (tan) including strawberries, cherries, apples, sweetberry honeysuckle, potatoes, wheat, and corn. The Hokkaido Agricultural Research Center is located on the southern edge of the city (yellow rectangular pattern) and works to create improved agricultural techniques for cold climates. The bright pink dots on the left of the image are highly reflective snow patches on the mountaintops.

Bukhara, Uzbekistan
Landsat OLI/TIRS
1:288,000

An oasis on the Silk Road trading route, Bukhara (gray) has a long agricultural history. Today, most of its surrounding land is pasture and desert (beige), but crops (red) are an important economic contribution from the region. Vegetation is very reflective in the shortwave (near) infrared wavelength. In this image, this wavelength is represented by red, so vegetation is highly visible.

Raqqa, Syria
Landsat OLI/TIRS
1:155,000

The image is bisected by the Euphrates River flowing across the bottom third of the image, and the region is part of the "cereal belt" of Syria, producing the vast majority of wheat in the country. After the dam in the bottom left of the image was built in 1968, the population of Raqqa (gray) increased and with it local cultivation (red).

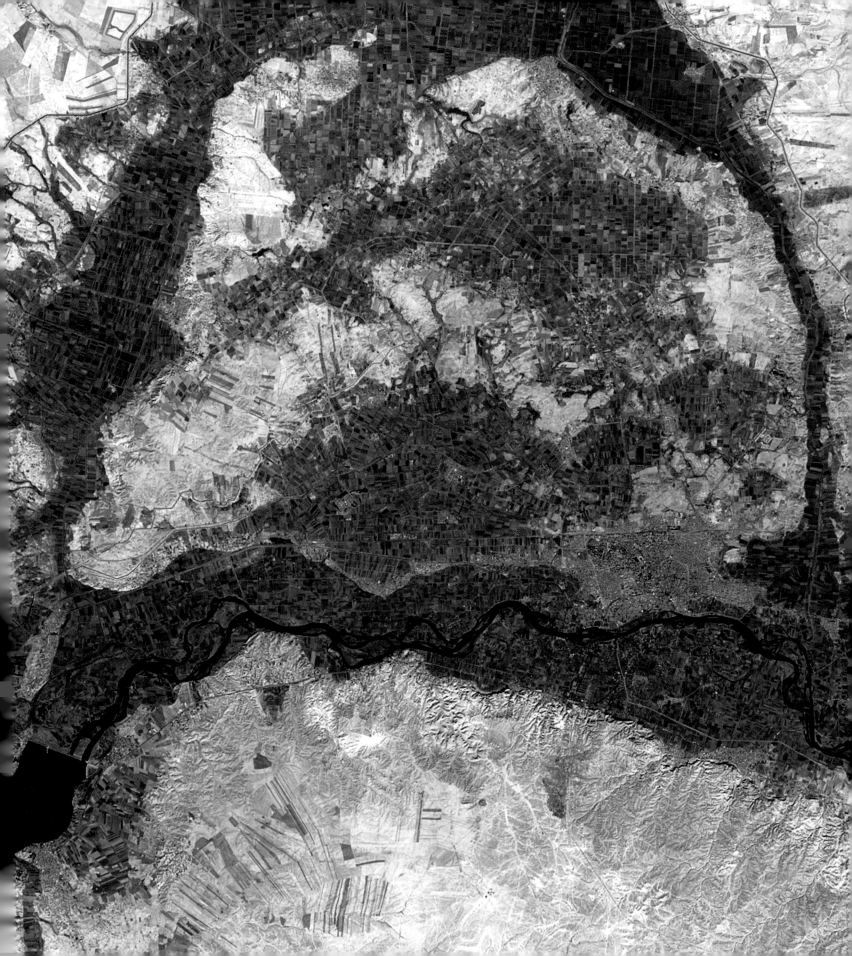

Dundee, Scotland
Landsat OLI/TIRS
1:180,000

Dundee sits on the mouth of the River Tay flowing through the center of the image and historically was the home to about sixty jute mills where vegetable fiber was mixed with whale oil to make sacks and carpet backing near the end of the nineteenth century. The city is pictured in purple on the northern banks of the River Tay near the center of the image. City building materials (such as concrete) and bare soil (like that found in fallow agricultural fields) have a similar reflectance and are often captured similarly by satellites. This is why both the city itself and the empty fields appear purple in the image. Fields with actively growing crops appear in different shades of green.

Pierre, South Dakota, United States
ASTER
1:125,000

According to the South Dakota Department of Agriculture, 98 percent of farms in the state are family owned and operated, and over 2,500 farms have been in the same family for over one hundred years. This ownership pattern is not typical for the majority of the United States. Much of the agricultural land is being turned over to larger conglomerate land owners. Route 14 appears as a diagonal blue line dividing the top half of the image.

Quebec City, Quebec, Canada
Landsat OLI/TIRS
1:364,000

Quebec City (purple) abuts the Saint Lawrence River, which bisects the image diagonally. Long, narrow agriculture plots are centered around both the river itself and smaller tributaries, used to irrigate the surrounding agricultural land. The long agricultural plots, also called long lots, seigneuries, or fiefs, are a remnant of French feudalism. The plots allow many different parcels to have river access for irrigation, while still keeping neighbors physically close. The arrangement also allows for a more equal division of soil types for each farm along the flood plain perpendicular to the river.

Delhi, India
Landsat OLI/TIRS
NDVI
1:675,000

Here, we see an NDVI image of Delhi and its
surrounding regions. NDVI mages are used to
highlight vegetation. Delhi and its surrounding satellite
cities (depicted in black in the image) are densely
surrounded by agriculturally vegetated land (the
white regions of the image). Two main satellite cities,
Meerut and Hapur (large black dots) appear to the
east. The transportation networks outside of Delhi
link smaller surrounding settlements, known as
periurban areas.

Delhi, India
Landsat OLI/TIRS
1:240,000

This image highlights the urban center of Delhi, India, and its surrounding highly agricultural land. Two satellite images were required to create the image: one image representing the surrounding crops' growing phase and the other depicting the surrounding crops' dormant phase. By essentially "subtracting" one image from the other, we created this composite image—which shows the change in vegetation cover between the two time periods. Blue regions of the image are most likely agricultural lands or seasonal vegetation. The dark blue regions indicate the areas with the greatest change in vegetation, while the yellow and orange regions undergo less vegetation change and could be urban centers with little vegetation (like Delhi, pictured in the entire bottom left quadrant of the image) or areas of consistent vegetation cover.

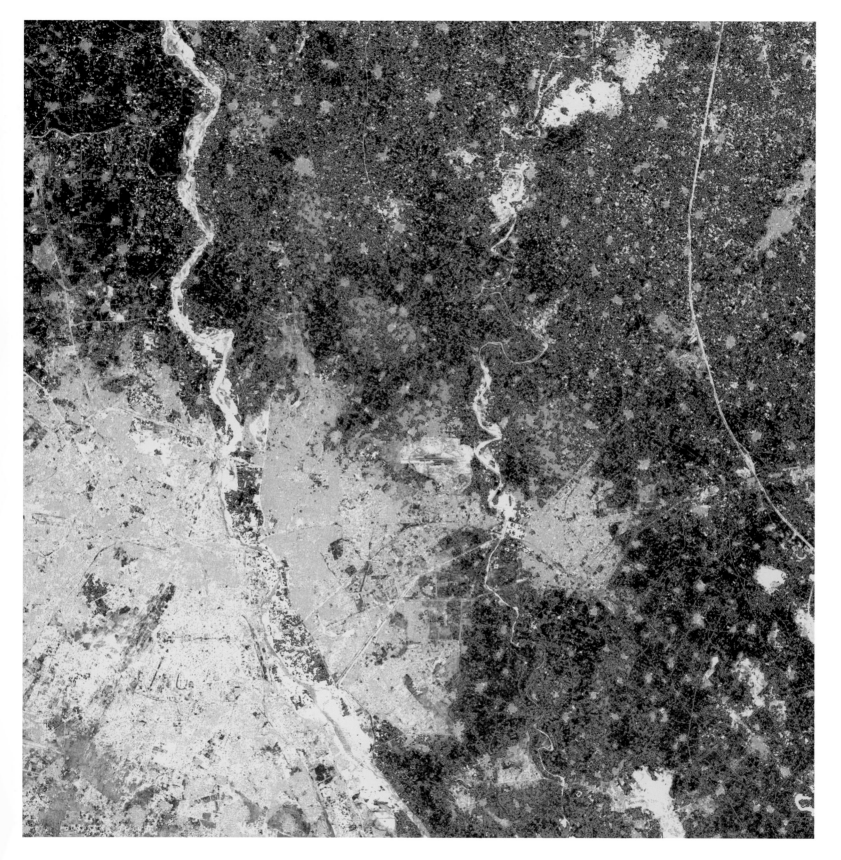

urban
imprints

borders
routes
plans

As a child, I was told that the Great Wall of China was the only human-made feature you could see from space. That's not true. Cities are the most visible and long-lasting human imprint on the land. How and where we build cities matters for the residents, for the environment, and for global sustainability. Modification of Earth's surface through urban development is fundamentally changing the way living organisms and systems interact with each other and our planet. Urbanization transforms land-scapes—from vegetated surfaces to buildings, pavement, sidewalks, and other artificial impervious surfaces. The process of city building requires resources often from afar and increas-ingly with materials from around the world. Cities themselves transform natural habitats. They break up habitats and transform local climates, rivers, water bodies, and air quality. The use of irrigation and major water diversion projects have allowed urban settlements to grow to immense sizes. From the design of neighborhoods and street layout, how and where urban areas develop affects resource use, biodiversity, human health, social cohesion, and ultimately, sustainability.

The impact of cities on the environment has expanded from local to global. As we move toward an urban century, a critical question is: What forms of urbanization are most environmentally and socially sustainable? Which forms of urban development promote stronger, more vibrant communities? What are the consequences of different patterns of urban development? What urban forms minimize use of resources such as water, energy, concrete, steel, and other raw materials? What are the social trade-offs of three billion new urban dwellers living in megacities of ten million rather than in smaller cities?

Not only does the imprint of urban development reveal urbanity's physical presence, it also gives a glimpse into our overall health. A city's transportation network has often been called the skeleton of a city and shapes its overall development. As a city becomes more spread out, people are less able to walk from origin to destination and must rely more heavily on cars. More compact development can improve human health through opportunities for walking and biking, for both recreation and transport. Public transit could help reduce congestion and transport energy use while improving human health and property values, but require minimum residential and employment densities to make them successful and well utilized.

This section focuses on three key physical expressions of urban areas. The administrative boundaries of a city also circumscribe an area's unique economic, development, and social-cultural history. Two neighboring towns may share a similar geography but, due to differences in preferences, architectural styles, or available building materials, may develop quite differently. The first of the three subsections,

"Borders," highlights these physical differences. The second subsection, "Routes", underscores the importance of transportation networks in defining the spine and subsequent form of an urban area. The third subsection, "Plans", celebrates the creativity and aspirations of different city visions.

At this critical time in history, we have an opportunity to improve the way we envision, design, and develop cities for a healthier urban humanity and the planet.

borders

borders

When the bus hit one of the many bumps on the dusty dirt roads, my knees slid across the seat's cold metal back and into its rigid frame. We sat three adults, packed together on seats made (nearly forty years earlier) to fit two American schoolchildren. Of course, that did not account for the chickens and other animals. The luggage rack on top of the bus was also overflowing—not with luggage, but with more people. The most striking thing about crossing the border was that it was nearly seamless; we barely paused to stamp our passports. We were still in a Spanish-speaking country. We could hear the woman at the bus station shouting, "Venta pupusas caliente!" The only evidence that we had crossed the border and were now in a different country was the smooth, paved roads lined by neat and tidy houses.

El Salvador–Nicaragua border
October 1991

For thousands of years, border towns have coexisted, often in an interconnected, mutually beneficial relationship. In modern times, discrepancies in economic development and political structures that exist on two different sides of a boundary are increasingly common. Even in a united European Union, differences in architectural styles, cultures, and lifestyles that evolved over generations endure despite a common currency. Although borders may be fluid, such as within the European Union or sub-nationally, between states or provinces, they still highlight the differences in histories and past experiences.

Border towns may share parallel histories, but their futures often diverge. Sometimes, the economic disparities between two border towns are so stark that they may as well be thousands, not just a few, miles apart. A town on one side may be more rural or agrarian in nature, with few municipal services and even less infrastructure. The town on the other side of the border may be highly developed, wealthy, and modern. One side may lack political institutions and legitimate governance such that it gives rise to illegal activities and a black economy. In many places, the disparity in wealth and development is a reflection of significant differences in the scale of foreign investments and the engagement and commitment of the local government.

These differences are magnified as we observe border towns from space. The striking contrasts reflect not only differences in the built environments, streetscapes, and building materials, but also undergird the stark differences in cultures.

Mexicali, Baja California, Mexico / Calexico, California, United States
Landsat OLI/TIRS
1:234,000

The border between the United States and Mexico is clearly visible about one-third of the way down the image. The twin cities of Mexicali, Mexico and Calexico, California, United States (purple, south and north of the border), are archetypal examples of the "urban pileup" effect. Many Mexicans move to cities on the U.S. border. There are fourteen cross-border cities and towns with populations greater than 15,000 people. Except for San Diego–Tijuana, all of them have a significantly higher population on the Mexican side. Strong agricultural roots can be seen on both sides in the mosaic of fields. In some ways, the urban and rural characteristics of each country are opposites. Calexico features larger, linearly arranged agricultural plots and a smaller, less defined urban footprint. Mexicali has a larger urban center with wide, linear boulevards, surrounded by an irregular patchwork of agricultural plots.

**Buffalo, New York, United States /
Niagara Falls, Ontario, Canada**
Landsat OLI/TIRS
1:220,000

An international tourist destination, Niagara Falls is the busiest United States–Canada border crossing, with close to four thousand people crossing each day. Niagara Falls appears slightly left of the image's center in light blue. Developed areas appear in bright yellow. Niagara Falls, New York, is located on the right, and Niagara Falls, Ontario, on the left of the Falls. To the southeast, Buffalo, New York, is clearly visible along the coast of Lake Erie. To the northwest, St. Catherine's, Ontario, borders Lake Ontario. The yellow-orange rectangular shapes concentrated in the upper right and lower left quadrants of the image are agricultural plots.

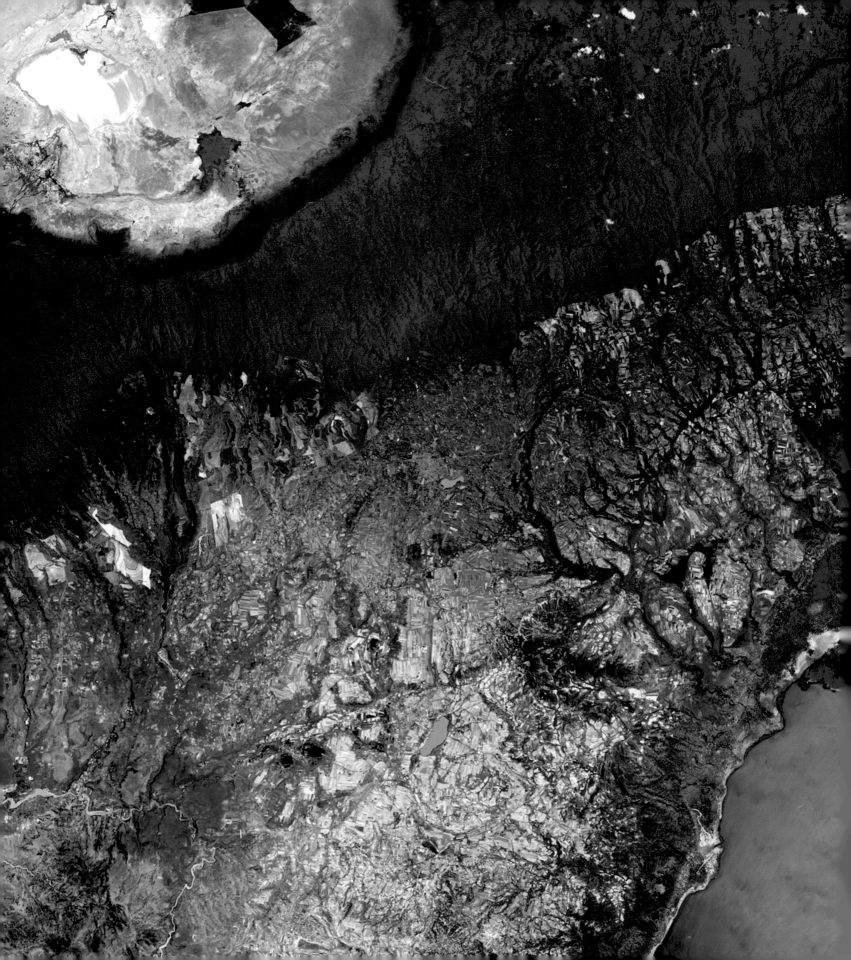

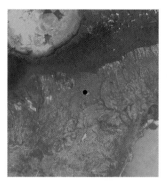

Karatu, Tanzania
Landsat OLI/TIRS
1:185,000

This is an image of many borders, natural and artificial. In the upper left corner of the image, we can see the Ngorongoro Crater, formed after a volcano collapsed in upon itself some 2.5 million years ago. This crater is ten to twelve miles (sixteen to nineteen kilometers) across. Along its rim, the stark red region shows a heavily vegetated area, home to protected wildlife and a popular safari destination. The city of Karatu is barely visible compared with such grand features—a small, linear blue-green region in the center of the image. In the lower right corner, we see part of Lake Manyara, home to one of the largest colonies of flamingos in Africa. During the dry season, the width of the lake decreases by nearly half due to lack of precipitation in the region.

**El Paso, Texas, United States /
Ciudad Juárez, Chihuahua, Mexico**
Landsat OLI/TIRS
1:271,000

Two cities appear in purple along the Mexican and
U.S. border. El Paso, Texas, is concentrated between
the border and the Franklin Mountains (in dark green).
Ciudad Juárez, Chihuahua, sprawls south of the
border. Agricultural lands in red spiral outward along
the path of the Rio Grande, the nearest water source.
Three border-crossing pedestrian bridges link the
two distinctive cities.

**Brazzaville, Republic of the Congo /
Kinshasa, Democratic Republic of the Congo**
Landsat OLI/TIRS
1:220,000

Though separated by less than two miles (about three kilometers) at the widest point across the Congo River, Brazzaville, in the Republic of the Congo (light blue bordering the north of the river), and Kinshasa, Democratic Republic of the Congo (light blue bordering the south of the river), are not yet connected by road or bridge. Boats are the primary source of transportation between the two urban centers.

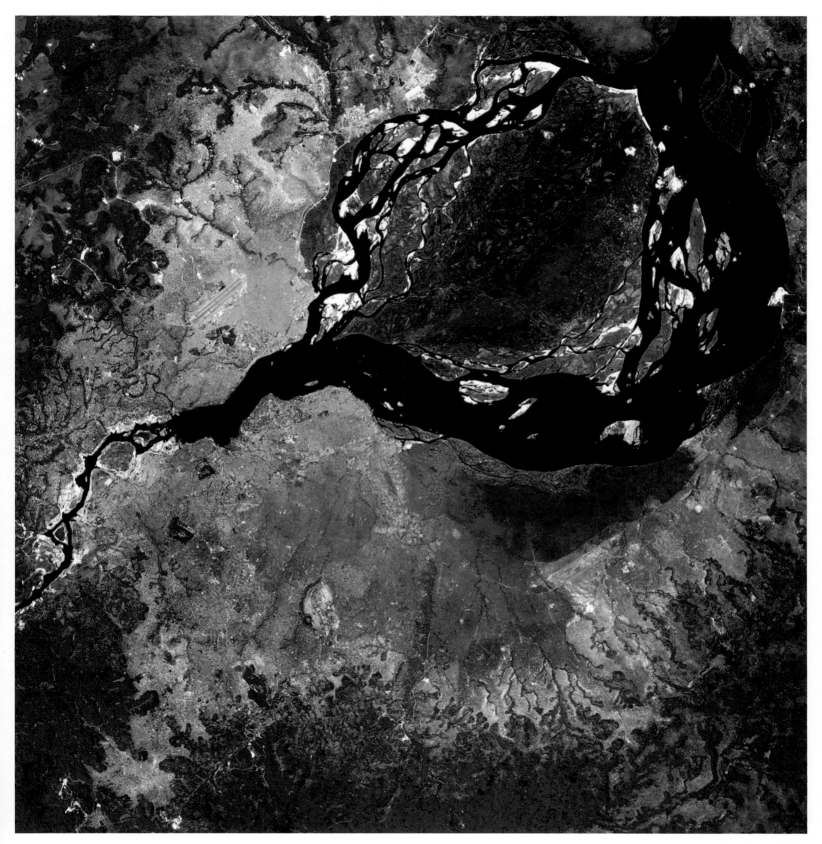

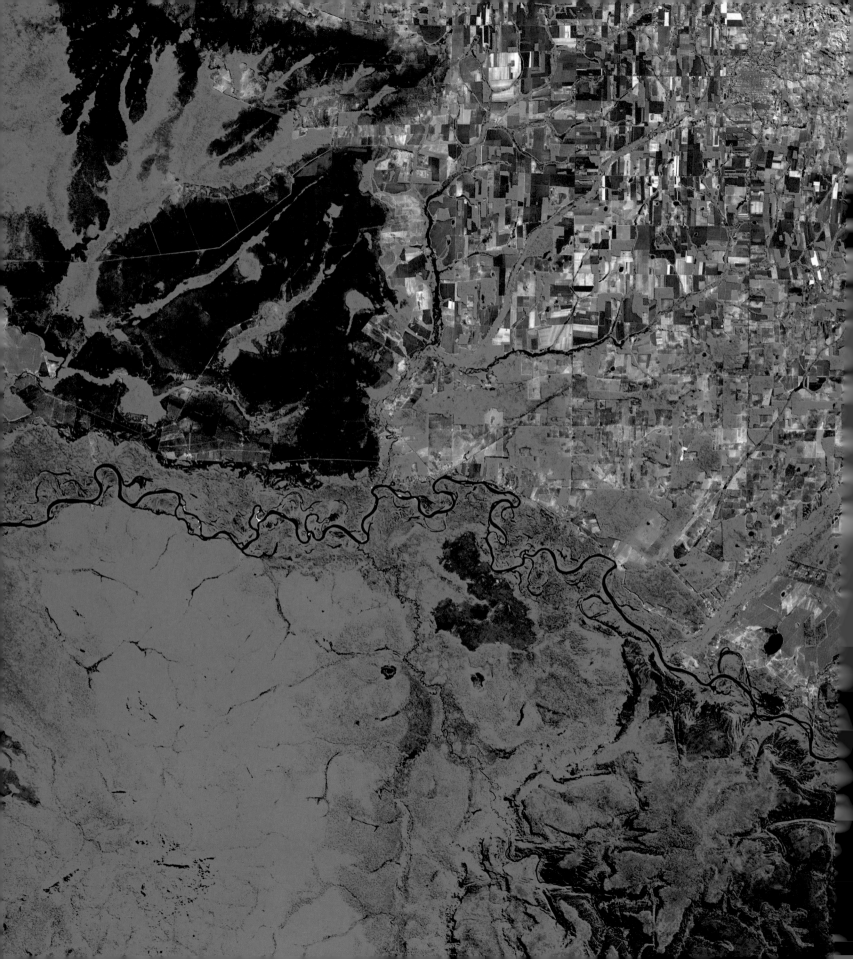

Pimenteiras do Oeste, Brazil / Parque Nacional Noel Kempff, Bolivia
Landsat OLI/TIRS
1:320,000

Borders often separate distinctly different land uses on opposing sides. This is the case in the Brazil (north)–Bolivia (south) border pictured here. The border follows along the Rio Itenez O Guapore, which roughly bisects the image. The intensive agricultural production of Brazil (pink) contrasts the open forest of the Parque Nacional Noel Kempff in Bolivia (green). Two settlements are pictured on the Brazilian side of the border—the smaller Pimenteiras do Oeste just north of the border, and the larger Cerejeiras—about 1 mile (1.6 kilometers) across—near the top-right of the image. These settlements appear in pink because they have a similar reflectance to the bare soil in agricultural plots.

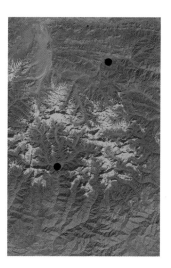

Namche Bazaar, Nepal / Zhaxizongxiang, China (Tibet)
Landsat OLI/TIRS
1:625,000

Mount Everest sits in the center of the image, dividing Tibet (China) from Nepal. To the north is the small settlement of Zhaxizongxiang—a village on the road heading to the Everest base camp on the Tibetan side, which branches off Route 318 running along the top of the image. To the south is the area of Namche Bazaar—located in the Khumbu region, often known as the gateway to the Himalayas for tourists and trekkers. Urban centers in the Himalaya region are expanding, fueled by migration from rural areas, valleys, and plains; the growth of religious, ecological, and adventure tourism; and recent social unrest. In 1981, less than 10 percent of the Himalayan population lived in a town or city. By 2000, the urban population in the region had doubled. The region is also prone to natural hazards such as earthquakes, fires, landslides, and floods, increasing the vulnerability of these remote mountain settlements. The highly vegetated regions of Nepal (red) south of Mount Everest contrast with the dry Tibetan Plateau (brown), with an average elevation of over 14,800 feet (4,500 meters) to the north.

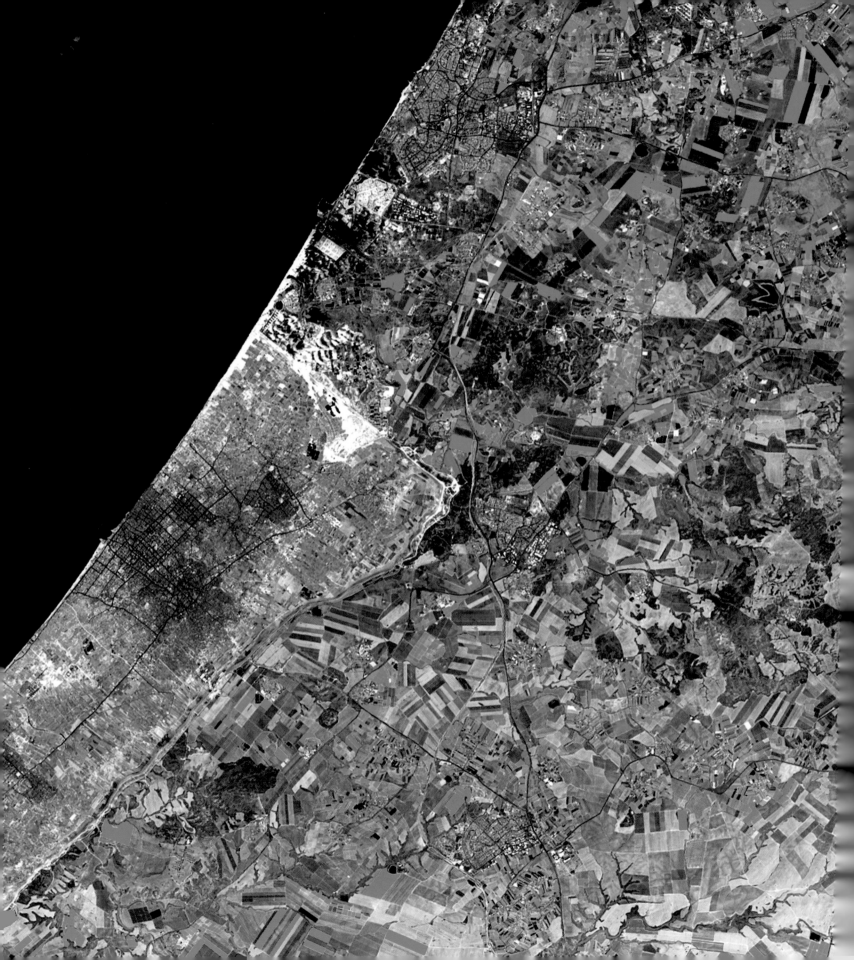

Gaza City, Palestine / Ashkelon, Israel
ASTER
1:145,000

A conspicuous border divices the two coastal cities of Gaza City, Palestine (south), and Ashkelon, Israel (north), in the left portion of the image. Ashkelon is surrounded by an abundance of rectangular plots in various orientations. Gaza City, dominated by urban development (in purple), is strained for open space.

Pyongyang, Democratic People's Republic of Korea / Seoul, Republic of Korea (next two pages)
Left: SRTM DEM (NASA Shuttle Radar Topographic Mission Digital Elevation Model)
1:8,300,000

Pyongyang, Democratic People's Republic of Korea / Seoul, Republic of Korea
Right: VIIRS (Visible Infrared Imaging Radiometer Suite)
1:8,300,000
The left image shows elevation of the region for reference; white regions indicate land with higher elevations. Elevation models are often used in remote sensing to correct for distortions in light due to the varying terrain in mountainous regions. On the opposite page, the bright nighttime lights of Seoul are captured in the center of the image, in contrast to the near total darkness of North Korea. North Korea's capital Pyongyang appears just to the northwest of Seoul and north of the faint east-west line of lights along the border. Pyongyang continues to be plagued with bouts of electricity shortages despite multiple stations servicing the city. The city of Shenyang, China, is most visible in the upper left quadrant of the image, while Shanghai, China, is visible in the lower left quadrant and the southwestern region of Japan occupies the bottom right quadrant of the image. The white and black regions of the two images are often reversed over the land. Larger cities, indicated by bright nighttime lights in the right image, are usually located at lower elevations, indicated by black regions in the left image.

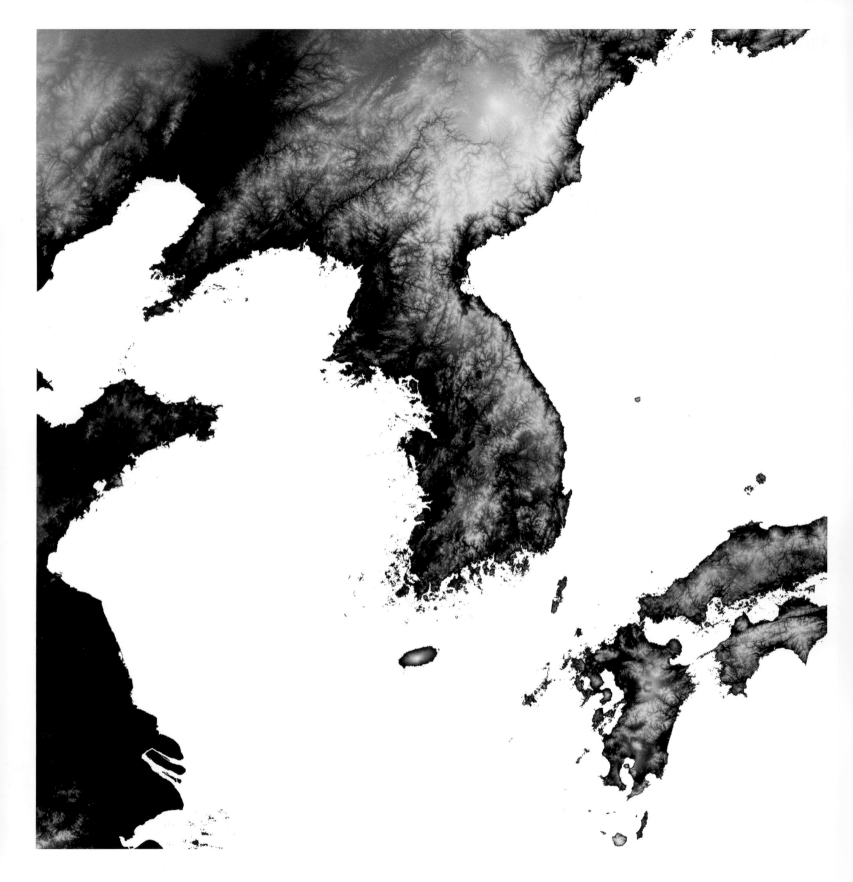

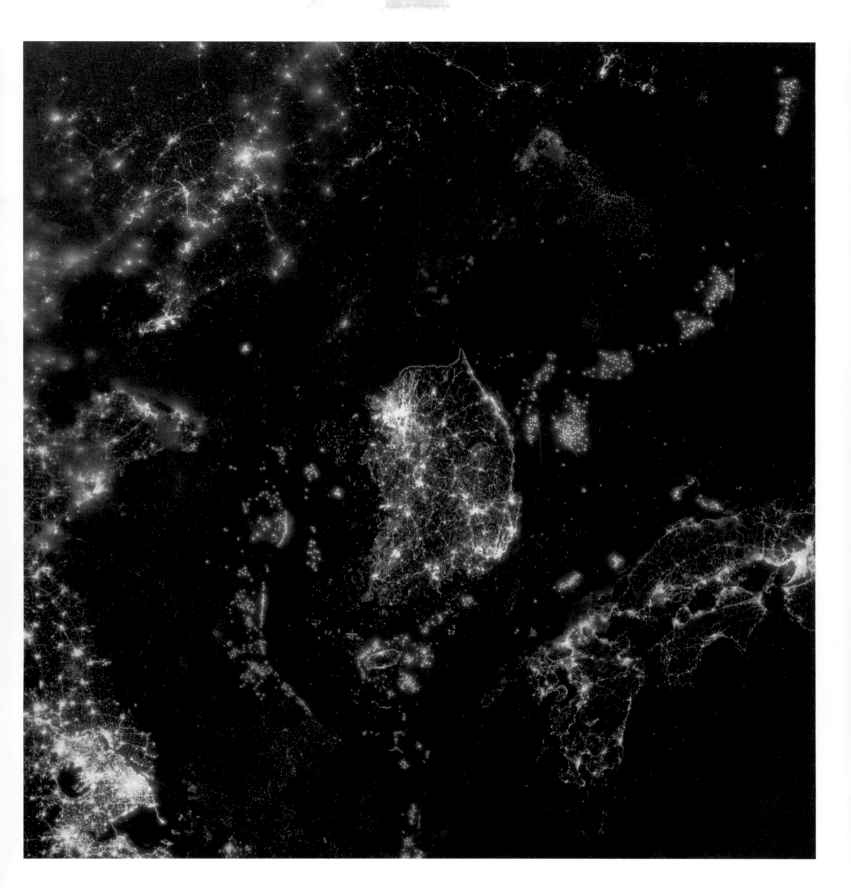

routes

routes

It's spring. We have now ridden through three seasons, and summer is quick to show its grip. Everyone tells us that it is hotter this year than the last. A pattern. With the power cut late in the afternoon, our hosts decide it is too hot to stay indoors for dinner. Wheeling out a motorcycle and a scooter from the entryway, the five of us—split two and three—board the cycles in search of a brighter night. The city is dark as we leave their house, but within minutes, rows of dazzling streetlamps cast sharp shadows before being swallowed into our previous path. The air is cool as it covers our faces, adding a deeper film of visibility to the beaming streets. New shopping malls and sprawling theaters line both sides of the horn-filled highway. As we inch toward the city center on the now winding narrow paths, crowds of buyers and sellers flock the busy streets—shouting promises of new opportunities, styles, and perhaps for some—a way of life.

Surat, India
March 28, 2007

Some argue that the Roman Empire constructed the first roads; others point to the Indus Valley civilization in the northwest regions of South Asia. Regardless of who built them first, there is no doubt that roads are central to modern civilization: they enable societies to transport people, materials, products, and services between what otherwise would be disconnected and often distant places.

The phrase "the other side of the tracks" illustrates the power of transport infrastructure to separate and divide communities. Highways through cities, especially, have often created economic and racial divides that persist for generations.

Around the world, roads can take many forms and reflect the diversity of size, ownership, speed of travel, width, length, shape, and mode of transport, among other attributes: interstates, city streets, blocks, cul-de-sacs, lanes, paths, alleys, highways, parkways, turnpikes, avenues, and boulevards. They range from grid-like patterns of city blocks to sinuous cliff-hugging roads that follow coastlines to interconnected roundabouts. From space, we can see ribbons of transportation networks that celebrate human-kind's ingenuity and freedom from the limitations of bipedal transport. Transportation networks, be they railways, six-lane high-speed highways, airport runways, or one-way residential streets, have significant impacts on ecology and society. Roads displace animals and fragment wildlife corridors, which can lead to isolated populations that become vulnerable to demise.

The size and orientation of city streets can either promote or discourage walkability, connect or divide a community. How we compose our roads has the potential to connect more than just two buildings, regions, or cities; roads have the possibility to act as a bridge between communities.

Kuwait City, Kuwait
Landsat OLI/TIRS
1:194,000
Kuwait City, derived from the Arabic word for fort, *kūt*, is located below the Kuwait Bay of the Persian Gulf in the center of the image. Minoprio, Spencely and Macfarlane, a British firm, created a master plan in 1952 with hopes of transforming what began as a mud-walled settlement roughly five square miles (thirteen square kilometers) in size to "the best planned and most socially progressive city in the Middle East." Using post–World War II New Town urban design along with Garden City movement principles developed by Ebenezer Howard, such as low-density neighborhoods and ring roads radiating from the city center meant to avoid congestion and disease of older cities, this city layout has led to a dependency on cars.

Amsterdam, Netherlands
Landsat OLI/TIRS
1:177,000

A city of cyclists and canals, Amsterdam is known for its canal ring. *Grachtengordel* in local parlance, this word describes the city's interconnected network of waterways located in the center of the image, just south of the Amstel River. Elaborate and connected cycle paths throughout the city trace intricate canals, as cyclists rule the roads in what is often known as the "bicycle capital of the world." In the 1950s and 1960s, cyclists were beginning to be pushed out by an increasing number of motorists, but by the beginning of the twentieth century the number of bikes surpassed the number of cars, and the city has never looked back.

Houston, Texas, United States
Landsat OLI/TIRS
1:445,000

Houston, Texas, is known as the only major U.S. city that lacks formal zoning. Though the city does employ regulatory design ordinances and limits certain land uses, it has spread out like a web from the urban center and is automobile dependent. A crack in the grid, the Buffalo Bayou, flows toward the city's center from the east. In 2017, Hurricane Harvey dumped more than fifty inches of rain onto the city in just a few days, turning roads into rivers and knocking bridges off their foundations.

Mendoza, Argentina
Landsat OLI/TIRS
1:147,000

Mendoza (light purple) s ts along the eastern foothills
of the Sierra de los Paramillos—a secondary range of
the Andes Mountains. The Mendoza River flows along
the bottom of the image, irrigating the desert oasis city
itself through an intricate system of irrigation ditches
or *acequias* parallel to the city's characteristically
wide tree-lined streets (dark purple). The region is
also known for its surrounding agricultural lands,
mainly vineyards.

Taipei, Taiwan
Landsat OLI/TIRS
1:151,000

The greater Taipei city region is actually formed by two cities—Taipei City located in the eastern portion of the city with a clear grid network, and New Taipei City to its southwest. Home to the indigenous group called the Ketagalan, Taipei's urban center development has occurred in multiple phases: first, by a migration of the Han Chinese in the early eighteenth century, and later, after Taipei's colonization by Japan in 1895. This second wave of expansion demolished Taipei's old city walls and developed expansive urban infrastructure such as roads and water drainage throughout the city.

Santiago, Chile
Landsat OLI/TIRS
1:160,000

Flanked by the Andes to the east, Santiago is the national railroad center and well connected to surrounding regions by roads (a large ring road is visible circling the city), highways, subway, and air. The triangular shape in the center of the image is the site of the original city limits bound by the Mapocho River and Santa Lucía Hill (not clearly visible in the image, but it rests inside the easternmost point of the triangle), which was used as a lookout in the early days of the city.

Doha, Qatar
Landsat OLI/TIRS
1:143,000

Three years after Doha was officially named Qatar's capital, Llewelyn Davies designed the city's first master plan in 1974. This plan consisted of a larger street grid with ring and radial roads often disconnected from smaller streets. To the east of the city, the Hamad International Airport is clearly visible—its (purple) runways aim into the Persian Gulf.

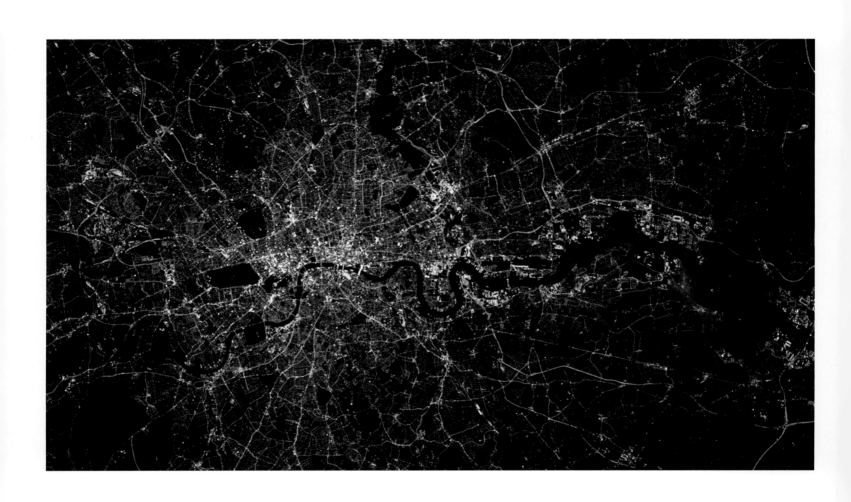

London, United Kingdom
ISS (International Space Station) image—digital camera
1:210,00

This image of London shows the different intensity
of lighting throughout this city bisected by the River
Thames. Traditional roadway lighting, historically
made from high-pressure sodium bulbs, tends to be
warmer in color—similar to the roadway lighting on
the outskirts of the city. The central city lighting tends
to reflect the cooler, bluer hues of LED or fluorescent
lightning. We can also see this fluorescent lighting on
the Tower Bridge, the bright yellow/white line crossing
the River Thames just left of the center of the image.

New York, New York, United States
Landsat OLI/TIRS
1:500,000

This image focuses on the navigable waterways surrounding this port city. Different wavelengths of light can penetrate water to different depths. The red wavelength of light only penetrates about five meters into a water body, but blue light can penetrate up to thirty meters depending on water clarity, and green light falls in between these two depths. This image uses the blue and green wavelengths of light to show shallow water depth. Deep water regions appear dark blue, while shallow water regions appear light blue. Shallow water regions are most visible near the shoreline and barrier islands south of Long Island and also throughout the Hudson River to the west of Manhattan, where boats are also visible moving through the waterway. The Port Newark and Elizabeth Port Authority Marine Terminal, located north of Staten Island on the west side of Newark Bay, is the busiest container terminal on the east coast of the United States, processing 3,602,508 cargo containers in 2016.

New York, New York, United States
ASTER
1:150,000

Thank the Commissioners' Plan of 1811 for New York City's navigable, gridded streets. Heralded as the "single most important document in New York City's development," it defined the original street layout between Houston Street and 155th Street. The image here also captures the abundance of piers on Manhattan's west side, the high density of tall buildings and their accompanying shadows (black in the image) just south of Central Park (bright red) and also in lower Manhattan, and the Brooklyn, Manhattan, and Williamsburg Bridges (west to east) on the south end of the island. Parks and other open space appear as bright red regions in the image.

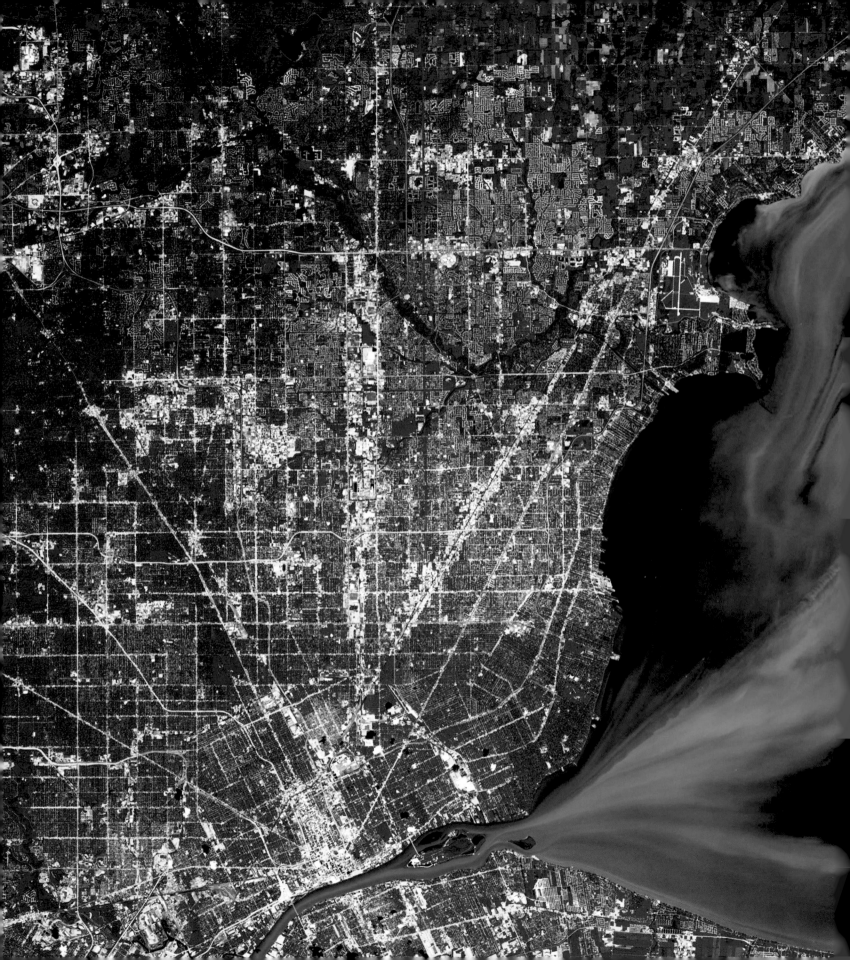

Detroit, Michigan, United States
Landsat OLI/TIRS
1:442,000

Given its history as the Motor City, Detroit's roads are a prominent feature of the city's transportation network. Hexagonal street patterns from the early nineteenth century remain in the downtown area of the city. Lake St. Clair dominates the right half of the image in green, but it is actually quite shallow, with an average depth of only 11 feet (about 3.3 meters). The largest delta in the Great Lakes can be seen in the northeast corner of the lake, where the St. Clair River enters the water body flowing from Lake Huron. The reflectance of water appears in darker green using this light combination, while the water's shallow depth and heavy sediment loads appear lighter in color.

plans

Our new house was about thirty miles from our old apartment, located in a new housing development in a city known only to NASCAR fans. Thankfully, we never heard the sounds of the Speedway—we were too far away and it closed only two years after we moved in. We could hear the freeway. All night long, semi-trucks with their Jake brakes howled and stuttered as they sped down the highway.

When we lived in the apartment complex, I could only ride my bike back and forth in the parking lot. In our new neighborhood, everything was new and sparkling. There was a network of streets I could explore. One day I rode my bike out of our development and into nearby streets. Every street parallel to ours was identical—with the same houses and the same cul-de-sacs. All the houses were different shades of beige stucco with tidy lawns in the front. It was thrilling to be part of something neat and orderly. We had achieved the American Dream.

Ontario, California, United States
June 1978

Every great city starts with a vision. For centuries, urban or town planning has been a principal means to organize, structure, and standardize infrastructure, dwellings, and life within a city. Street design, city walls, and urban layouts have been developed for military purposes, to safeguard security, and to protect against invasion. Over time, street configurations, grid plans, and zoning emerged as central methods to limit access to roads and neighborhoods and to protect residents from vehicles or fast-moving traffic, or simply to reduce the amount of vehicular movement near residential zones.

Whereas historically, cities were places where housing and commerce were tightly woven and closely located, today many cities are zoned for single uses, leading to the separation of places of employment from places of residence. There is growing recognition that many tenets of twentieth-century urban planning and zoning have led to high consumption levels of land, water, and fossil fuels, affecting energy and climate systems worldwide.

Urban growth patterns evident in the United States during the second half of the twentieth century have now been adopted by many Asian developing nations. While the spread of American-style single-family detached homes in the suburbs is one relatively small dimension of urban expansion in the early twenty-first century, the environmental impacts of this trend appear to be of a proportionately greater significance because of the per-capita energy and land consumption required to sustain low-density development. If the built environment of the urban United States contributes to the country's rank as the world's most significant polluter, then the adaptation of similar built environments in nations with much larger populations presents a pressing environmental problem. New ideas about urban planning are critical to create healthier, more sustainable cities for the future.

Cape Coral, Florida, United States
Landsat OLI/TIRS
1:183,000

Planned in 1957, Cape Coral, Florida, is shaped by an extensive canal system creating just under 400 miles (about 640 kilometers) of artificial coastline as the water flows through this interconnected waterway. Forcing water to flow through this extended waterway has affected tidal water levels, since ebbing and flowing water now must travel through this canal network. The Yucca Pens Unit State Wildlife Management Area, which sits north of the city (brown), is protected for hunting, fishing, and frogging.

Palmanova, Italy
WorldView, DigitalGlobe Image
1:10,000

Constructed in 1593 by the Venetians, Palmanova
features a nine-pointed "star fort" or *trace italienne*—
a defense system and geometric wonder. Six roads
radiate from a hexagonal piazza at the center, with
three leading to city gates and three to ramparts.
Four ring roads link these radial roads. The whole
fortress is surrounded by a moat, and the main
buildings inside the fortress face the piazza. Despite
its protected structure, no actual battles occurred
in the fortress. People were reluctant to inhabit the
fortress once it was completed, and it was later used
as a residence for exonerated criminals. Today, about
five thousand people live in the fortress, which is a
national monument.

Chandigarh, India
Landsat OLI/TIRS
1:250,000

Designed by renowned urban planner and architect Le Corbusier in 1950 to have an ordered, grid street network, Chandigarh, located at the foothills of the Himalayas, was the first planned city in India after independence in 1947. The structured linear street network and large reinforced concrete buildings is atypical for Indian cities. The structure of the city has been likened to a living human body, with streets serving as "blood vessels," homes as "cells," and cars and people as "blood."

Brasília, Brazil
CBERS
1:264,000

Influenced by the work of Swiss-French urban planner Le Corbusier, the architect Lúcio Costa won a design competition for the new capital of Brasília, which was built between 1956 and 1960 to represent a new, more equitable capital in the interior of Brazil. It has been criticized for its large-scale design, which is car-dependent and is often cited as missing a sense of place. The city plan is also critiqued for its lack of mixed-use zoning, which integrates different land use types such as residential, commercial, industrial, and cultural. The bird's-eye view of the city is often compared to the blueprint of an airplane, with the residential and commercial sections of the city spread out like wings from the main "fuselage" made up of the Eixo Monumental (Monumental Axis) and its iconic grand administrative structures.

La Plata, Argentina
CBERS
1:87,000

Heralded as South America's first planned city and a turning point in Argentinian urban planning, La Plata was designed by engineer Pedro Benoit in the form of a perfect thirty-six-by-thirty-six block square grid with a gothic cathedral at the city's center and city squares or roundabouts at major intersections every six blocks. Diagonal boulevards transect the space to create a unique star-shaped pattern from above—especially at night, since La Plata was the first urban street grid in Latin America to use electric street lighting.

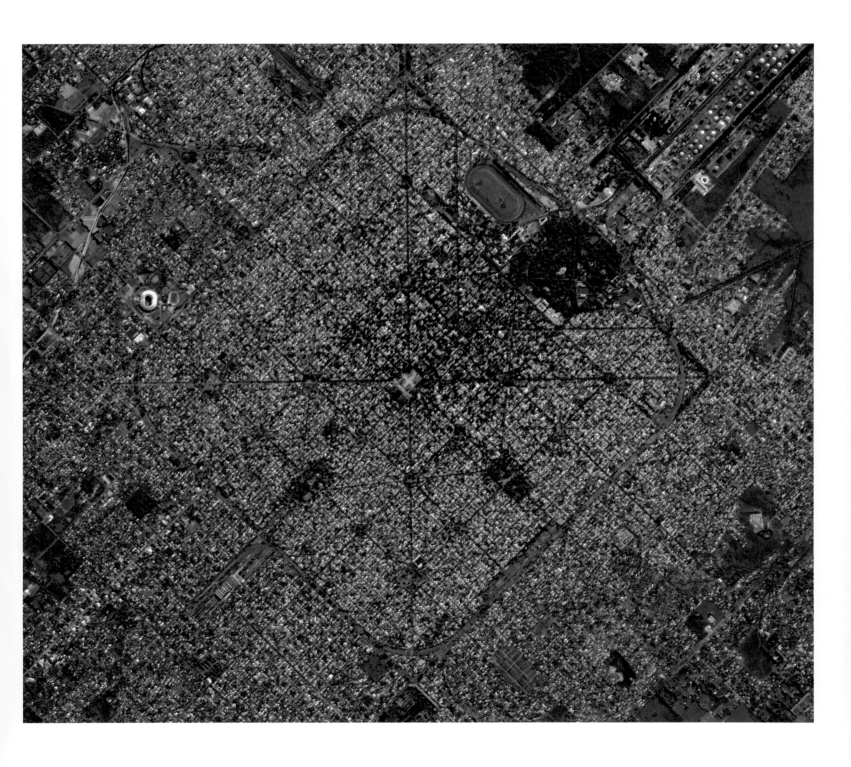

Siem Reap, Cambodia
ASTER
1:150,000

An important cultural, religious, archaeological, architectural, and artistic center, Angkor Wat spreads over 400 square kilometers (about 155 square miles) and was the center of the Khmer Kingdom (802 CE–1432 CE). Home to extensive stone temples and waterways including canals, dikes, and reservoirs, such as the West Baray—a reservoir 5 miles (8 kilometers) long by 1.5 miles (2.4 kilometers) wide with an average depth of 2.5 feet (4 meters) pictured in the center of the image. It is the largest human-made lake of the Khmer civilization and was thought to originally have been filled by rainwater, although today a series of channels and moats leads to the water body. Monsoons can flood the ancient city, as pictured here in November 2002. We can also see the low water level in the baray in the bottom image, captured in February 2004. February is one of the driest months of the year, evidenced by sparser vegetation (red) in the bottom image.

Barcelona, Spain
World View, DigitalGlobe Image
1:14,000

The word *eixample* means expansion or extension
in Catalan, and the Eixample district was conceived
in the mid-1800s in response to Barcelona's high
rates of morbidity and mortality. The district's orderly
chamfered octagonal blocks with wide boulevards and
airy sunlit courtyards are in stark contrast to the dense
labyrinth of narrow streets of the older city located in
the top part of the image.

Nahalal, Israel
ASTER
1:80,000

Based on the principles of the kibbutz such as communal ownership of land, equality, and social justice, while still allowing for private ownership of land, Nahalal was developed by architect Richard Kaufmann in 1921 as a cooperative agricultural settlement. In a layout resembling the spokes of a wheel, shared housing and community buildings such as schools, shops, and administrative offices are located in hubs, and independently owned farmland radiates outward.

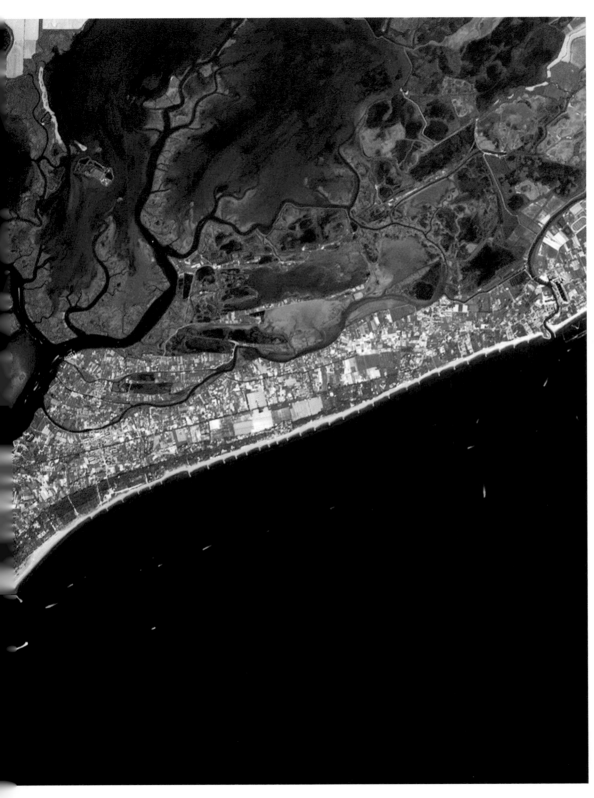

Venice, Italy
ASTER
1:105,000

The Ponte della Libertà is the road bridge connecting the historical center of Venice (green), made up of 118 small islands within the larger Venetian lagoon, linked by canals or bridges to the Italian mainland to the west. Barrier islands and marshes can be seen in the eastern portion of the image, separated by the Faro di San Nicolò (St. Nicolas Lighthouse) and the Punta Sabbioni Leuchtturm flanking the western and eastern edges of the waterway opening to the Adriatic Sea. Floating along the meandering canals of the historic city center, glimpses of waterfront palazzos, small neighborhood churches, and grand cathedrals are revealed to visitors and locals alike.

Beijing, China
Landsat OLI/TIRS
1:130,000

The focal point of the image is the Forbidden City (rectangular region in the upper left quadrant of the image). Urban or built-up areas appear in gold in the image, water bodies in black, and vegetated regions in blue. Some light yellow regions represent fallow/bare soil or particular building materials. The ring roads surrounding Beijing and the connecting linear road networks (dark gold/brown) illustrate movement within the city boundaries, as these roads link commercial centers and markets. They also illustrate how the urban network expands outward, enveloping the surrounding agricultural land, altering the land use of these regions.

transforming the planet

resources
expansion
vulnerability

Urbanization, sustainable development, and a healthy planet are inextricably linked. The world's urban population is expected to increase by about 170,000 people a day through the end of this century. The growth in urban areas involves the construction of buildings and roads, water and sanitation facilities, and energy and transport systems that are transforming the p.anet and cities worldwide. Recent estimates suggest that between $25 trillion and $30 trillion will be spent on infrastructure worldwide by 2030, with $100 billion a year in China alone. Designing, developing, constructing, and operating cities will require massive amounts of natural and human resources and result in substantive energy and carbon costs. Thus, there is an urgent need for much, if not all, of future urbanization, densification, expansion, urban revitalization, and all manifestations of the built environment to be low-carbon, energy-efficient, and of healthier design. As we continue to innovate and design smarter buildings and vehicles, it's very likely that we will rely even more on rare earth metals, copper, and other materials that are essential for improved technological performance.

Urbanization is transforming the global landscape, but it is also altering the exposure of millions of people to risks and hazards. Urban areas are vulnerable to multiple stressors—climate change, sea level rise, water shortages, droughts—and the frequency and magnitude of these are estimated to increase and intensify over the coming years.

This third section focuses on how an urbanizing humanity is changing the planet in places distant from city centers, and how a dynamic earth is transforming the urban condition.

resources

resources

My grandfather's mind remained sharp throughout his ninety-nine years. He told me stories of the ten cents a day he would earn to go down the shaft following the mule cart. Orphaned at age thirteen, he started working in the mines to take his father's place. Stories of how some bully (he could remember his name, but I can't) would try to take his ten cents each afternoon and how he would hide the dime under his tongue until one day he accidentally swallowed it and decided he better come up with a more solid plan. Of the kind man and coworker who soon realized that my grandfather clearly didn't have enough food to eat and would have his wife make him an extra lunch every day and who would care for him as part of their family. Stories of the cramped tunnel he would work in, where his arms could barely fit over his head as he picked away at the coal. How he was buried alive twice. The blare of the siren that would sound in town whenever there was a cave-in. And how families would flock to the mine shaft to see if their loved ones would resurface. As he aged, he developed black lung and major arthritis in his shoulders and fingers. He couldn't lift his arms above his chest, and I still have a vivid image of his hands—beautiful and powerful, with twisted fingers at every knuckle. He ended up leaving the mines early—at age fifty—because a family business on my grandmother's side offered him the opportunity to live a life where he was less likely to be taken from his family too soon.

McAdoo, Eastern Pennsylvania United States
February 2017

The world is experiencing a magnitude of city building unlike during any other time in history. Each week, the world adds the equivalent of a city of a million people and this is expected to continue until 2050. After 2050, the rate of city building might actually increase, as India is expected to begin its urban transition only in the second half of this century. The amount of resources required to construct and operate these cities is tremendous. China used 6.6 gigatons of cement between 2011 and 2013, compared with 4.5 gigatons that United States used over the entire twentieth century. In the early 2010s, 75 percent of the world's cranes were in Shanghai and Beijing, building skyscrapers and new gleaming cities. The Burj Khalifa in the United Arab Emirates, the tallest building in the world, is five times taller than the Eiffel Tower and has concrete that weighs the equivalent of 100,000 elephants—over 1.5 million pounds of steel. Providing adequate public services, sanitation, electricity, and housing for the large numbers of people who lives in these cities stretches the imagination.

In the past, city size was limited by the availability of nearby, accessible resources. Today, many of the raw materials to build cities are mined, excavated, processed, and transported from distant places. While the enduring image of a mining town may be the frontier of the American West, today's resource towns are spread throughout the world.

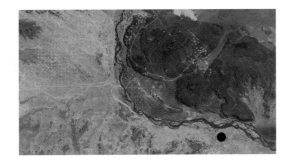

Puesto Hernández, Argentina
ASTER
1:130,000
Oil pads

Puesto Hernández is home to one of the oldest productive oil fields in Argentina, spanning an area of nearly 60 square miles (150 square kilometers). The oil field accesses the shale-gas reserves of Vaca Muerta, meaning "dead cow." The white tessellated pattern throughout the image is formed by the oil pads themselves, while the city (red urban vegetation and a black linear airport runway to the east) is located south of the river in the bottom right corner of the image.

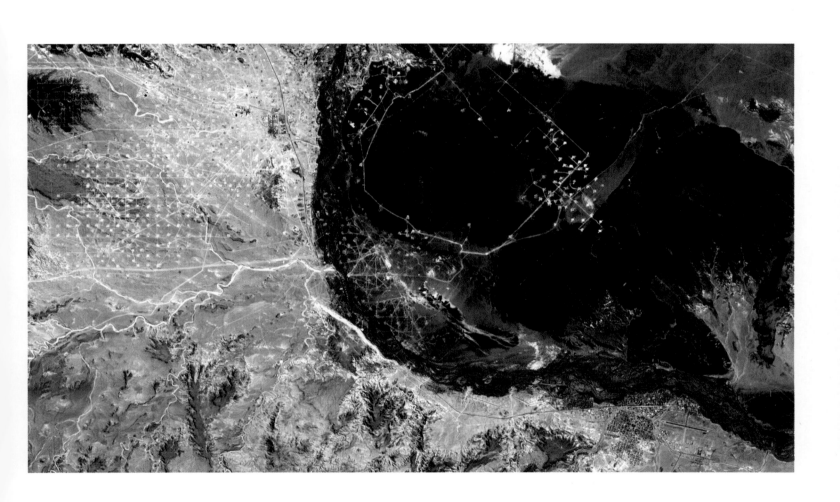

Kalgoorlie, Australia
Landsat OLI/TIRS
1:213,000
Gold mine

The town of Kalgoorlie, pictured in the purple grid-like pattern in the center of the image, is home to the one of the largest open pit mines in the world. The "Super Pitt" appears in blue, adjacent to the city itself (purple); the region is particularly rich in gold deposits. White Flag Lake is located in the upper left quadrant of the image. Because of gold's effectiveness and efficiency as a conductor, small amounts are often used in many electronic devices that power modern life.

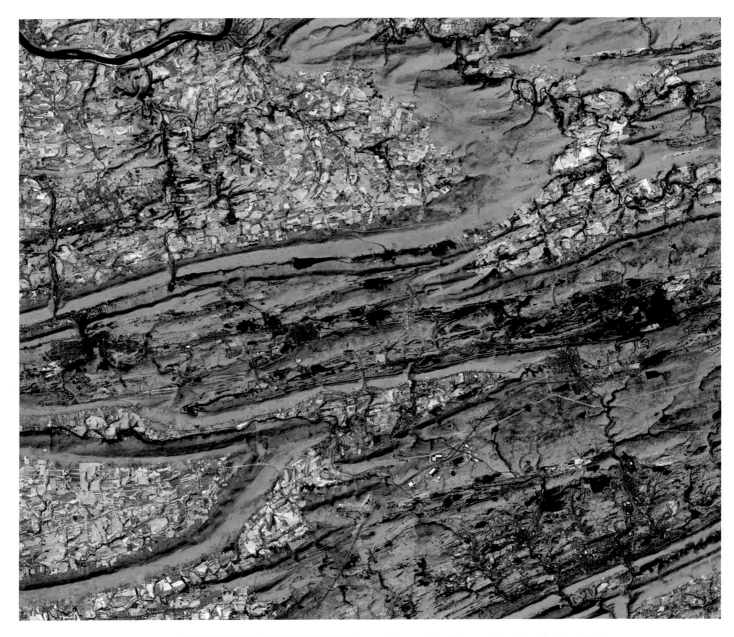

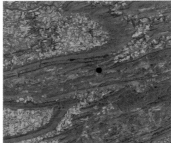

Centralia, Pennsylvania, United States
Landsat OLI/TIRS
1:222,000
Coal mine

Centralia, Pennsylvania, was a coal-mining town. Now it is a ghost town. An underground coal fire started in 1962 and is still smoldering. In the 1980s the U.S. government paid to relocate residents from carbon monoxide gas–infested properties. The town itself is located in light purple in the right central portion of the image, which was taken on April 24, 2014. To the

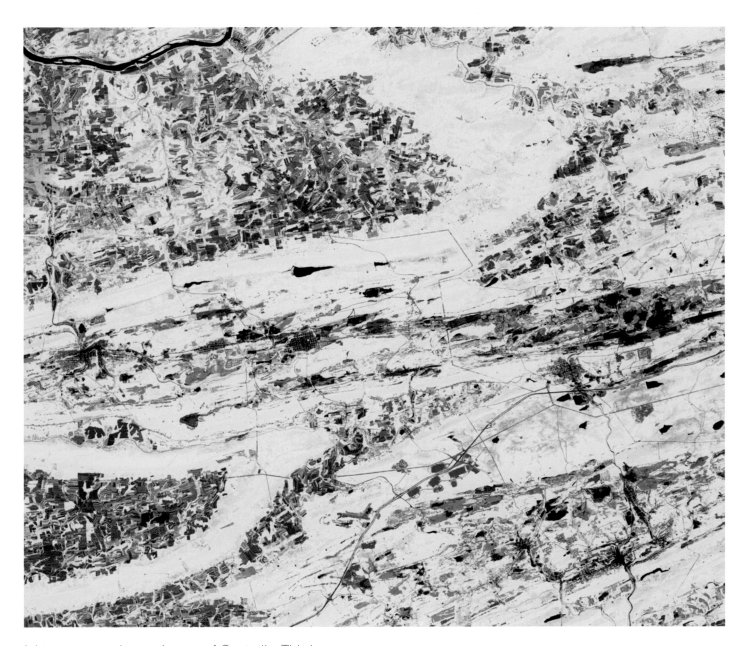

right, we see a change image of Centralia. This image
was created by using two images from the month of
May—one from 1985 and the other from 2014. Red
areas indicate regions where vegetation decreased
from 1985 to 2014, green regions show a vegetation
increase, yellow regions show where vegetation stayed
the same, while black regions had little vegetation in
both images. Looking closely at the change image,
the town of Centralia now appears green, indicating
an increase in vegetation due to the abandonment
of the town.

Rockhampton, Australia
Landsat OLI/TIRS
1:171,000
Gold mine

After the discovery of a gold deposit at Mount Morgan in 1882, Rockhampton was formed in 1902. The mine officially closed in 1981. Today, Rockhampton is known as Australia's beef capital and is home to over two million cattle within a 155-mile (about 250-kilometer) radius around the city.

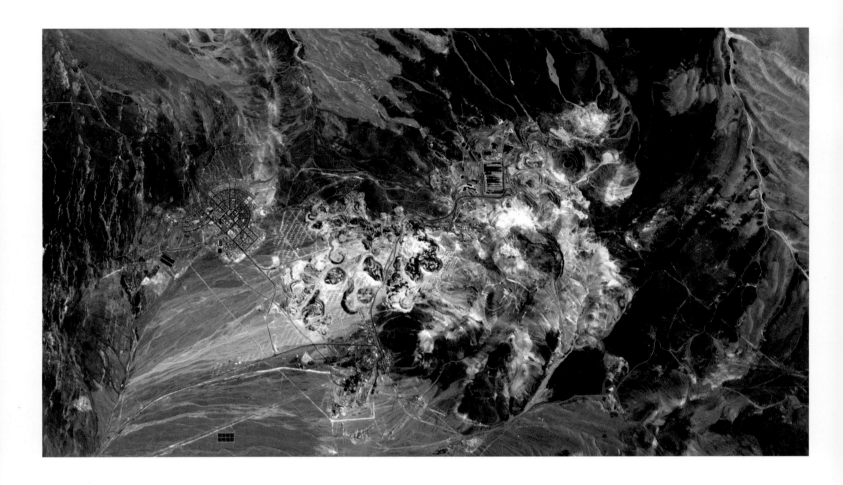

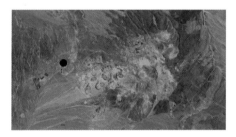

El Salvador, Chile
CBERS
1:93,000
Copper mine

El Salvador (The Savior) mine is a copper mine complex consisting of two open pit mines and one underground mine, while the city itself, pictured in the left portion of the image, has a unique arched shape and extra-wide roadways to accommodate mining trucks and equipment. The city is located at an elevation of nearly 8,000 feet (about 2,400 meters). Copper is often used in electrical devices, such as wiring and motors, in bustling urban centers around the world.

Broken Hill, Australia
Landsat OLI/TIRS
1:193,000
Silver, lead, and zinc mines

In 1883, silver and lead were found at Broken Hill (blue), a sub-arid region in southeastern Australia. Broken Hill was included on the National Heritage List in 2015 due to its status as a unique desert frontier town. The city sits atop a long, broad deposit over three miles in length. Zinc is used both as an alloy and as an oxide and is currently the fourth most commonly used metal in the world, following iron, aluminum, and copper. Its anticorrosive properties are critical for the galvanizing process to prevent rusting where a thin layer of zinc is bonded to the surface of other metals such as iron or steel. Zinc oxide is also used in rubber manufacturing.

Paramaribo, Suriname
Landsat OLI/TIRS
1:244,000
Wood

A former Dutch colonial town, Paramaribo is formed by long agricultural plots—most likely remnants of what's called "fishbone" deforestation where paths are cut into forested areas (shown in red) to remove timber. These fishbone patterns serve as larger reminders of the resources neeced to create the region's Dutch-inspired wooden architecture. Today, plans to construct tall timber skyscrapers continue to advance. Compared with other tall building materials such as steel and concrete, wood is a renewable resource and produces a significantly lighter weight structure. Tall timber construction also shows promise to mitigate global warming through reducing carbon emissions when harvested from sustainably managed forests. In fact, some even say the age of timber has officially begun.

Arequipa, Peru
Landsat OLI/TIRS
1:325,000
Open-pit copper and molybdenum mine

Mina Cerro Verde, established in 1993, is an open-pit copper and molybdenum mine pictured in blue at the bottom of the image just to the south of the city itself. Three volcanoes—Misti, Chachani, and Volcán Chachani—border the city to the north. Molybdenum is commonly used to make metal alloys—such as the steel used in engines and tools—to increase strength and conductivity and reduce corrosion.

Jharia, India (page 192—top left)
Landsat OLI/TIRS
February 23, 2017
1:475,000
Coal mine

Jharia is home to a massive coal fire that ignited in 1916 and has been burning ever since. The city itself is quite small and indiscernible in the image; it is located midway between the bright blue lake and the sickle-shaped water body at the top of the image. The mine and broad burning area spans the top of the image, and is especially concentrated in the upper right quadrant. It is estimated that over 37 million tons of coal have been lost to the flames, while other mineable areas are now unreachable. Attempts to extinguish the flames have been futile, since coal seam fires are extremely difficult to put out, especially because of their underground location and seemingly endless supply of fuel. Firefighters have tried using water and sand, and cutting off the oxygen supply, to no avail.

Jharia, India (page 192—bottom left)
Landsat OLI/TIRS
February 11, 1989 & February 23, 2017
difference image
1:475,000

This image shows regions where the temperature has changed from 1989 to 2017. Red areas show a decrease in temperature from 1989 to 2017, green regions show an increase in temperature over the same years, and in yellow regions the temperature remained relatively constant. These changes can also be inferred by visually comparing the temperature maps for the two years on the following page.

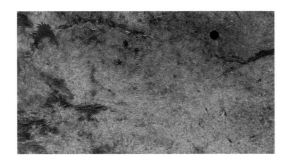

Jharia, India (page 193)
Landsat TM (top)
February 11, 1989
1:475,000
Coal mine

Landsat OLI/TIRS (bottom)
February 23, 2017
1:475,000
Coal mine

These two images show temperature for the region in 1989 and 2017. Yellow areas are warmest, while blue areas are coolest.

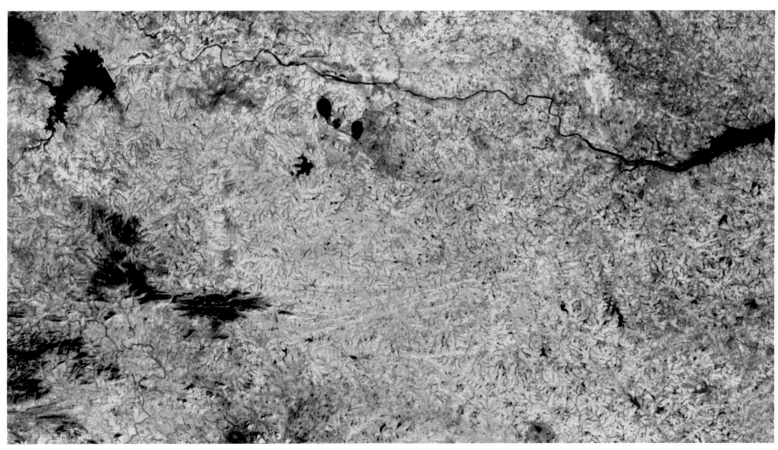

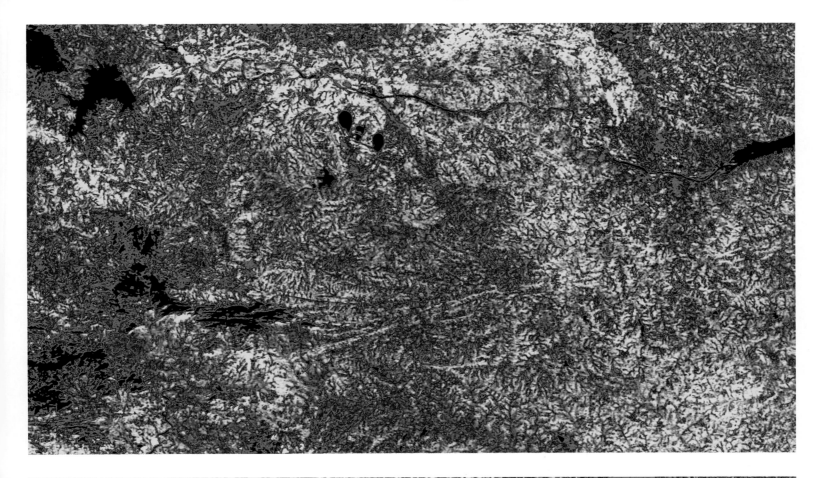

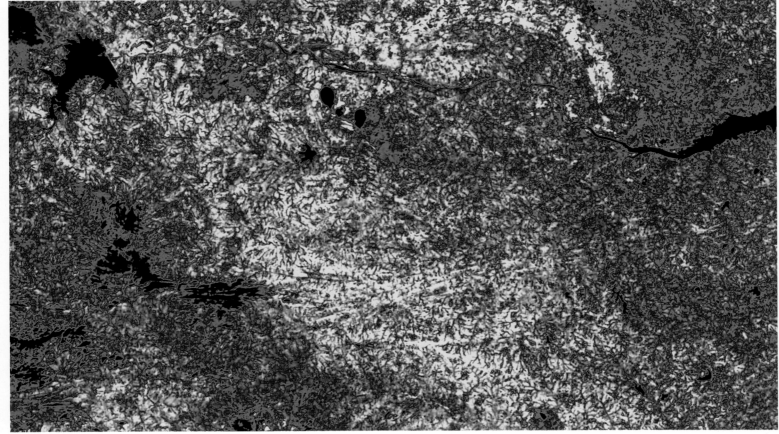

Djelfa, Algeria
ASTER
1:135,000
Salt mine

Djelfa is surrounded to the north and west by
formations pictured in black known as Salt Rock
(Rocher de Sel), which is formed by the erosion of
rock salts otherwise known as halite—the mineral
form of sodium chloride or table salt.

expansion

The driver turned around, looked at me with a skeptical eye, cigarette dangling from his lips, and said tersely, "There's nothing on the other side of those hills except rice paddies." Back in those days, my laptop hard drive couldn't hold a full satellite image, so we printed quadrants of the image on sheets of poster-sized paper. I looked at the printout, located the labeled latitude and longitude values, and compared these with the reading on my handheld GPS. There was no mistaking the new urban development on the other side of those verdant hills. I begged the driver to press on. We were on a dirt road with sugarcane fields lining both sides. As we reached the top of the hill, we could make out the shell of multiple three-story apartment buildings, land that had been razed for development, the outline of new roads. The driver abruptly stopped the van and got out, cigarette still dangling. He turned up to the sky and pointed, "Can they see what we're doing all the time? This wasn't here last week." Rome wasn't built in a day. I wonder how long it would take now.

Dongguan, China
November 1997

On every continent except Antarctica, urban landscapes are expanding into farms, forests, and rangelands; smaller urban settlements are merging into extended urban areas and enveloping smaller towns and villages. Walled cities were common from the fifth to the sixteenth centuries, and these barriers served to contain the growth of the city as well as physically separate the city from its environs. Contemporary cities have few, if any, material boundaries, and satellite imagery highlights that many are expanding unencumbered.

Contemporary urban expansion has characteristics that differentiate it from other periods in history: the scale of urbanization is extraordinary; cities are bigger than at any other time in history in terms of their physical extents. The urban extent of Tokyo-Yokohama covers 13,500 square kilometers, an area that is bigger than Jamaica (11,000 square kilometers). The rate at which land cover is being converted to urban uses is faster than during any other time in history, and much of this is taking place on prime agricultural land. In Puerto Rico, urban expansion resulted in the loss of one-quarter of the country's growing region.

The geographic center of rapid urbanization is also changing. Urban expansion in Europe and South America occurred in the 1950s through the 1970s. Urban expansion in the coming decades will take place primarily in Asia and Africa, transforming wildlife habitats and some of the most biologically diverse ecosystems in the world.

Lagos, Nigeria
Landsat TM
1:475,000

December 19, 1984
Population 3,291,000

Lagos, Nigeria
Landsat OLI/TIRS
1:475,000

December 18, 2015
Population 13,123,000

Lagos is being developed on top of coastal wetlands (creeks, swamps, lagoons) at a staggering rate. In 1881, the city covered only about 1.5 square miles (3.85 square kilometers) and has since expanded to a megacity (in green-blue) of over 1,400 square miles (2,700 square kilometers) in 2015. Although both images use the exact same wavelength combination to display the images, there are slight differences in color. Sometimes these differences can be due to seasonality or changing land uses. However, since the launch of the first Landsat satellite and sensor (Multispectral Scanner System) in 1972 to the current Landsat sensor (the eighth—cal.ed the Operation

Land Imager), the range of wavelengths the satellites and sensors are able to capture has greatly increased, from four in 1972 to eleven in 2013, significantly improving image quality.

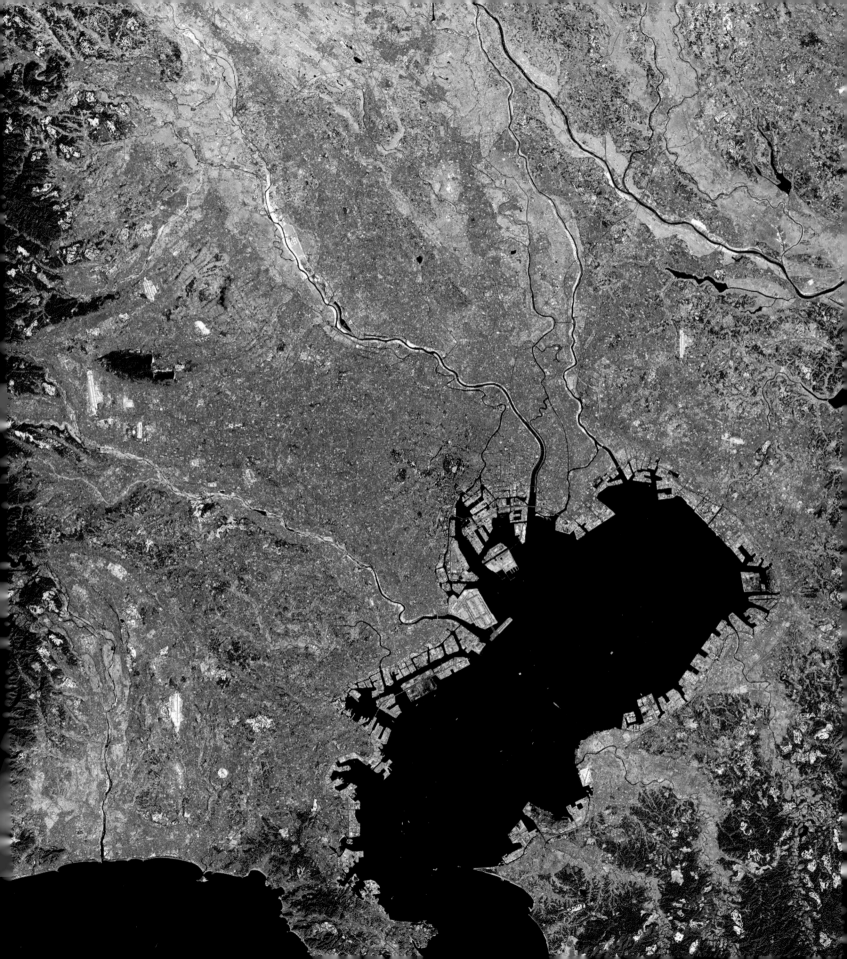

Tokyo, Japan (previous two pages)
Landsat MSS & Landsat OLI/TIRS
1:178,000

May 21, 1979—Population 28,169,000
February 16, 2017—Population 38,241,000

Tokyo is actually what is known as a "metropolitan pre-fecture" and is the only city in Japan with this distinc-tion. Essentially, it means the region combines features characteristic to both a city and a prefecture, which is similar to a state in the United States—a unit of govern-ment larger than a city or town. The Tokyo region has the highest population of any metropolitan area in the world—as it was home to more than thirty-eight million people in 2017.

Shenzhen, China
Landsat MSS, Landsat TM, Landsat OLI/TIRS
1:505,000

February 10, 1977—Population 43,000
February 7, 2016—Population 10,828,000

A small fishing village, Shenzhen's fate was forever transformed after it became China's first Special Eco-nomic Zone and the site of the country's experimen-tation with economic reforms. From 1978 to 2000, it experienced the highest rates of urban growth in the country. Rice fields and fish ponds became factories and skyscrapers. Today it is one of the most metro-politan and dynamic cities in China. Dramatic growth in the image is especially evident on Longxue Island on the eastern side of the Zhujiang River estuary in the center of the image and around Danan Mountain on the peninsula in the eastern part of the river in the second image. The improvement in satellite image resolution is also noticeable when comparing these two images.

Las Vegas, Nevada, United States
Landsat MSS
1:525,000

April 10, 1976
Population 345,000

Las Vegas, Nevada, United States
Landsat OLI/TIRS
1:525,000

October 12, 2015
Population 2,270,000

Las Vegas, Nevada, is one of the fastest-expanding urban areas in the United States. Not only can we see the expansion of the city itself in the western portion of the image, but the draining of Lake Mead to the east, emphasizing the level of resources cities and their inhabitants require for growth in delicate ecosystems.

Harbin, China
Landsat MSS
1:373,000

August 14, 1984
Population 2,180,000

Harbin, China
Landsat OLI/TIRS
1:373,000

September 20, 2015
Population 5,457,000

Sitting on the southern bank of the meandering Songhua River, the small fishing village of Harbin has expanded into one of the largest cities in northeast China. Its cold climate has made Harbin the home of the largest snow and ice festival in the world, known for its beautiful ice sculptures. The surrounding region has very dense agricultural plots, indicated by the bright red in the 2015 image. Pink regions in the 1984 image show areas of different vegetation cover—such as different crop types or growth stages of plants or grasses.

Nairobi, Kenya
Landsat TM
1:248,000

August 24, 1984
Population 1,040,000

Nairobi, Kenya
Landsat OLI/TIRS
1:248,000

February 25, 2016
Population 4,070,000

The Mombasa-Nairobi road has spawned extensive development in the city. In November 2015, a thirty-mile (fifty-kilometer) traffic backup consisting of over 1,500 trucks clogged the roadway approaching the city (pictured in light blue) and lasted for three days. Urban development (blue-green) is spreading northwest into the Karura Forest (red), yet approximately 2.5 million city dwellers live densely packed in the city's slums. Many rural residents move to the capital city to search for employment.

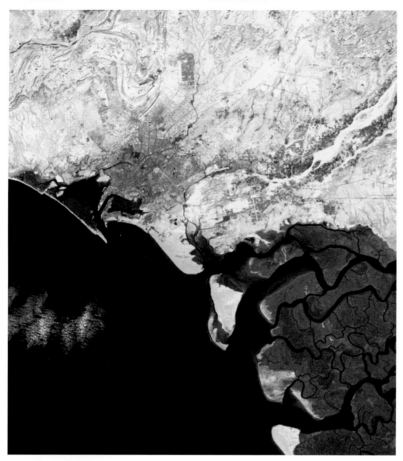

Karachi, Pakistan
Landsat MSS
1:477,000

October 16, 1972
Population 3,461,000

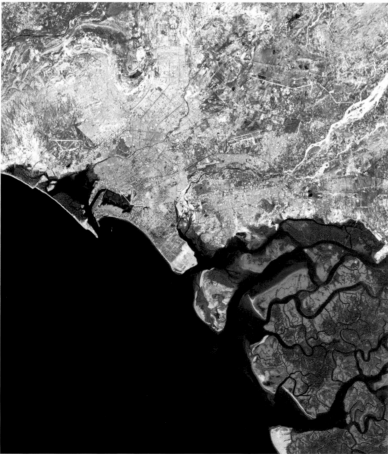

Karachi, Pakistan
Landsat TM
1:477,000

October 10, 1994
Population 8,185,000

Karachi, Pakistan
Landsat OLI/TIRS
1:477,000

October 20, 2015
Population 16,618,000

Karachi is home to Pakistan's industrial and financial center and is the most culturally diverse city in Pakistan. With its coastal location along the Arabian Sea, it is also a major port and transportation hub of the region. Southward growth is limited by the marshland (red) and Mahro Kotri Wildlife Sanctuary and Hawke's Bay to the southwest.

vulnerability

Suddenly, someone was shaking me vigorously, and I heard what sounded like thunder overhead. I didn't want to drop the dumpling from my soup spoon, and I was annoyed at being interrupted. Then I fell off my bed onto the floor and awoke from my dream. The violent shaking was an earthquake. I heard my mom yell for my four- and nine-year-old sisters. I tried to get to them, but the ground was shaking so hard that I couldn't stand up. It was like standing on top of a moving bowl of Jell-O. The thunder-like sounds were joined by deep rumbles from the earth's surface, punctuated by the terrible sound of our Southern California wood-frame house creaking and swaying. I was terrified. But just as quickly as it jolted me awake, it stopped. I could hear morning alarms going off and dogs barking. I realized then how vulnerable we are to the powers of nature.

Los Angeles, California
October 1, 1987

Humans build settlements in the most vulnerable places to inhabit: along coasts that are at risk of storm surges, hurricanes, and tsunamis; on the floor of deep valleys prone to flooding and mudslides; in tornado alleys; on steep precipices inclined to give way to landslides and avalanches.

Urban areas have always been vulnerable. The catastrophic eruption of Mount Vesuvius crushed Pompeii in 79 CE. The Yellow River flood in China killed about one million people in 1887. But cities have never experienced today's frequency of intense hazards. In the twenty years between 1980 and 2000, there were more natural disasters recorded than in all the years combined between 1900 and 1980. Climate change brings new risks to cities: sea level rise, drought, heat waves, water scarcity, extreme precipitation events, forest fires. A key question is: How can we build the cities of tomorrow to reduce their vulnerability and increase their climate resilience? How will we build, maintain, and service our cities when it's 125 degrees Fahrenheit (52 degrees Celsius) outside? Even asphalt melts at that temperature, as we learned from recent experience in India. Buildings constructed during the past century were probably not designed to withstand gale-force winds that accompany hurricanes, which are expected to increase in intensity and frequency in the future. And in a world where coping with hot weather means turning on the air conditioner, what will happen when extreme heat results in regular blackouts? We are reminded that despite human ingenuity, we are ultimately vulnerable to nature.

Miami, Florida, United States
Landsat OLI/TIRS
1:242,000

Due to its coastal location and elevation of only about 5 feet (1.5 meters) above sea level (compared with New York City at 33 feet—about 10 meters—above sea level), Miami is one of the United States' most susceptible cities to sea level rise due to climate change. A popular tourist destination, Miami's international airport can be seen in purple just to the south of the center of the image, and Dodge Island, both the city's port and cruise line hub, runs perpendicular to the barrier islands near the bottom of the image.

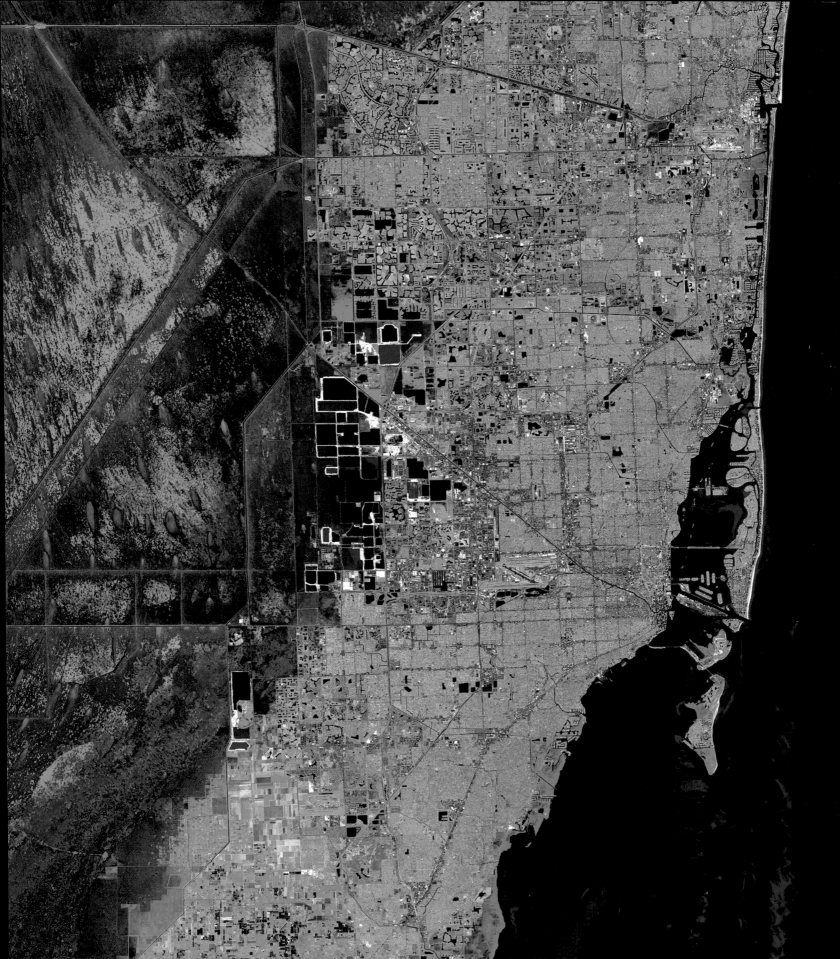

Joplin, Missouri, United States (right & following two pages)
Landsat TM & Landsat OLI/TIRS
June 26, 2009; July 2, 2011; & July 13, 2015
1:130,000

On May 22, 2011, Joplin, Missouri, experienced a tornado with winds over 200 miles per hour (300 kilometers per hour). The city was torn with a corridor of damage 1 mile (2 kilometers) wide and 6 miles (10 kilometers) long—shown as a smear of purple. Over 9,200 people were displaced. The damage allowed for the opportunity to redevelop and revive the city, including development of a variety of transportation options and improved stormwater management projects with assistance from the United States Environmental Protection Agency. It also encouraged the city to improve its disaster preparedness. Here, three images track the change— pre-tornado, one and a half months post-tornado, and four years into the recovery period. Regrowth can be seen in the third (2015) image.

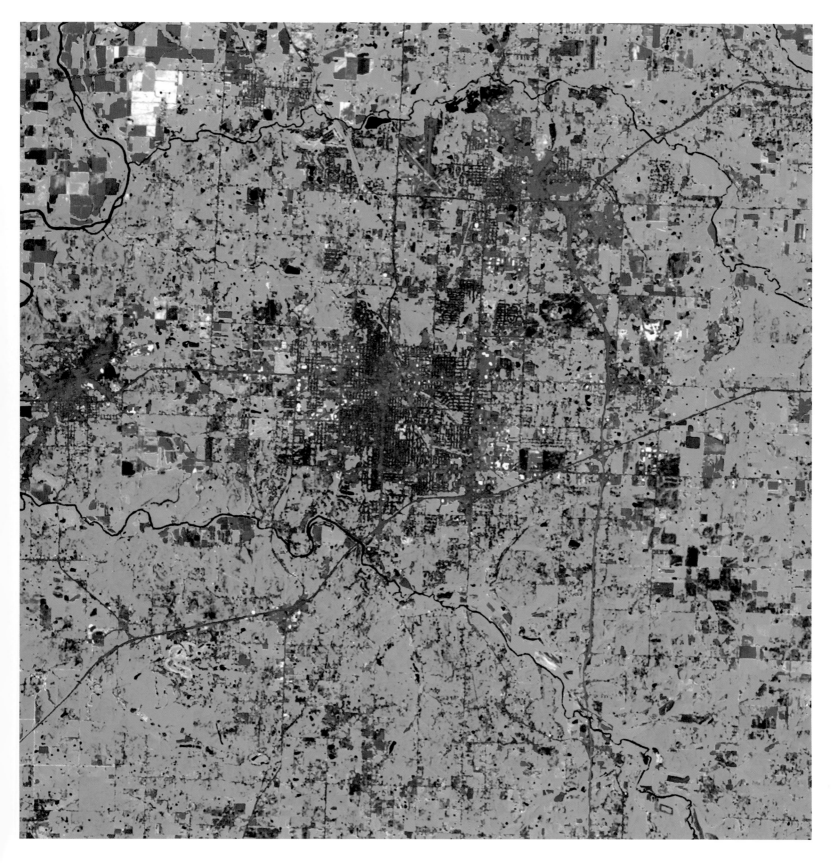

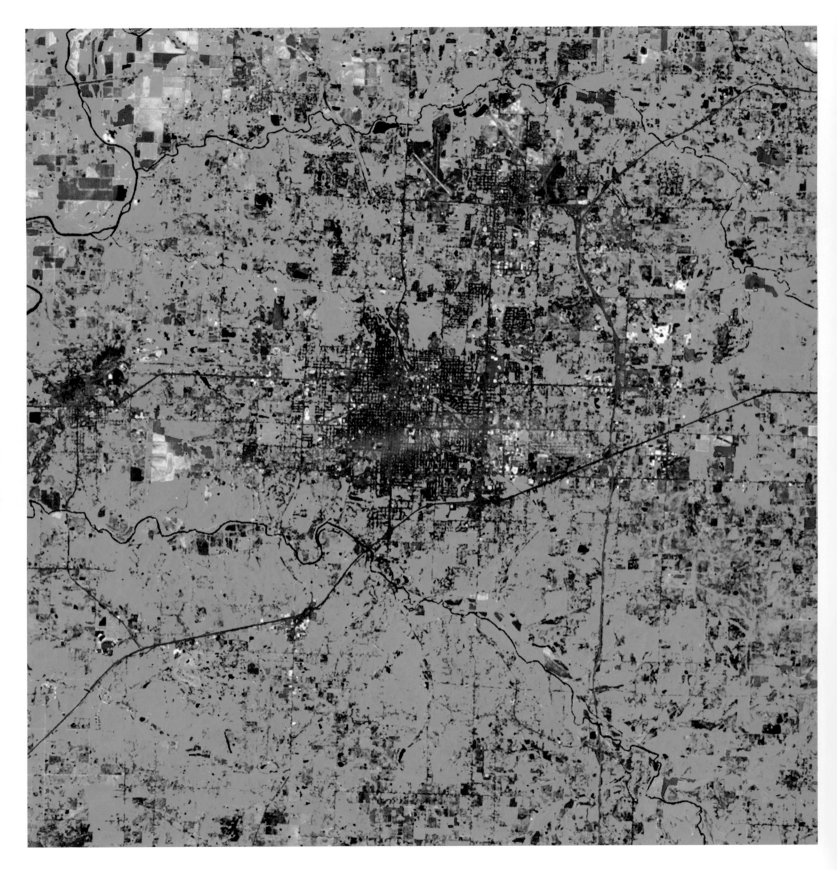

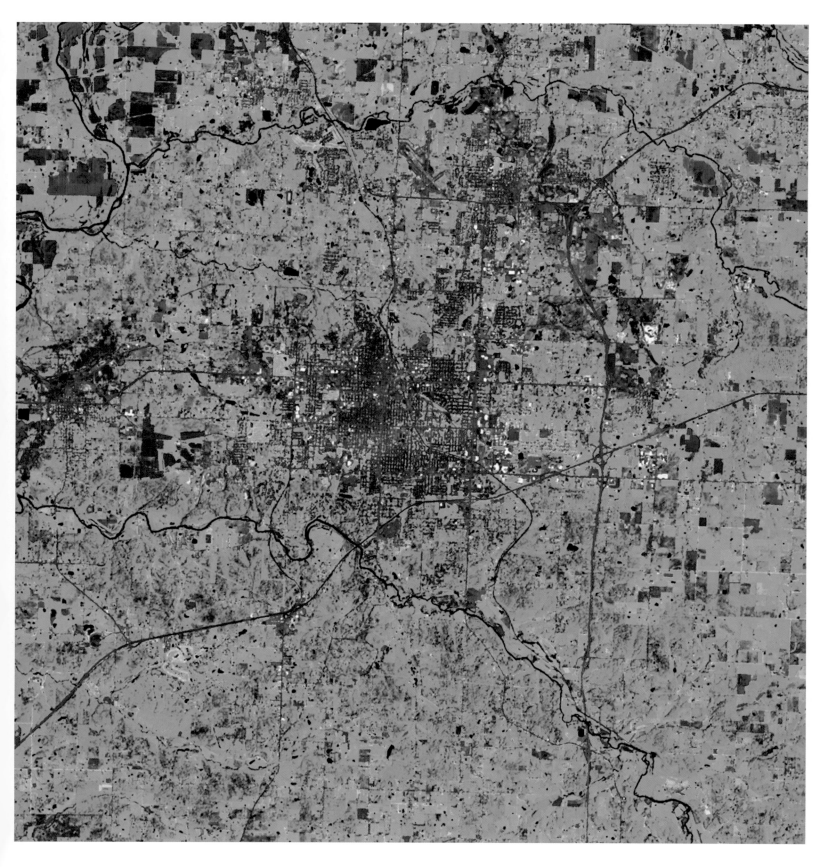

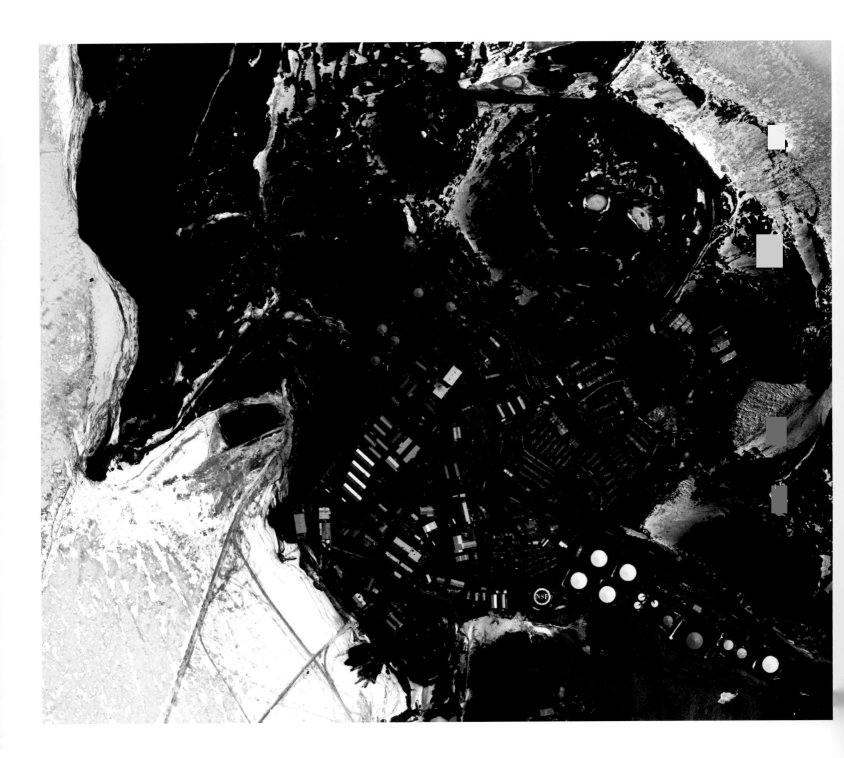

McMurdo Station, Antarctica
WorldView, DigitalGlobe Image
1:650

Antarctica's largest "city," McMurdo Station is about
2,500 miles (3,800 kilometers) south of Christchurch,
New Zealand, with an average annual temperature
of 0°F (-18°C). Summer temperatures can reach 46°F
(8°C) in summer and −58°F (−50°C) in winter. Multiple
buildings make up the "city," including administrative
buildings, stores, warehouses, dormitories, research
facilities, and a power plant.

Chittagong, Bangladesh
ASTER
1:160,000

Chittagong has experienced numerous floods and landslides over the years. Its low-lying average elevation of 130 feet (about 40 meters), the frequency of monsoons, and its location on the northern bank of the Karnaphuli River have resulted in sections of the city being flooded annually, especially from June to September. Due to the scarcity of water-handling infrastructure, such as drains and sewers, stagnant water abounds within the city limits during periods of heavy rainfall.

Timbuktu, Mali
Landsat OLI/TIRS
1:190,000

Sometimes called the "city of beige" and the "end of
the earth," Timbuktu blends into its surroundings at the
edge of the Kalahari Desert about eight miles (thirteen
kilometers) north of the Niger River with its sand-
covered streets and notable lack of cars. The Niger
River (blue) flows along the bottom of the image and
is surrounded by riparian vegetation pictured in red.

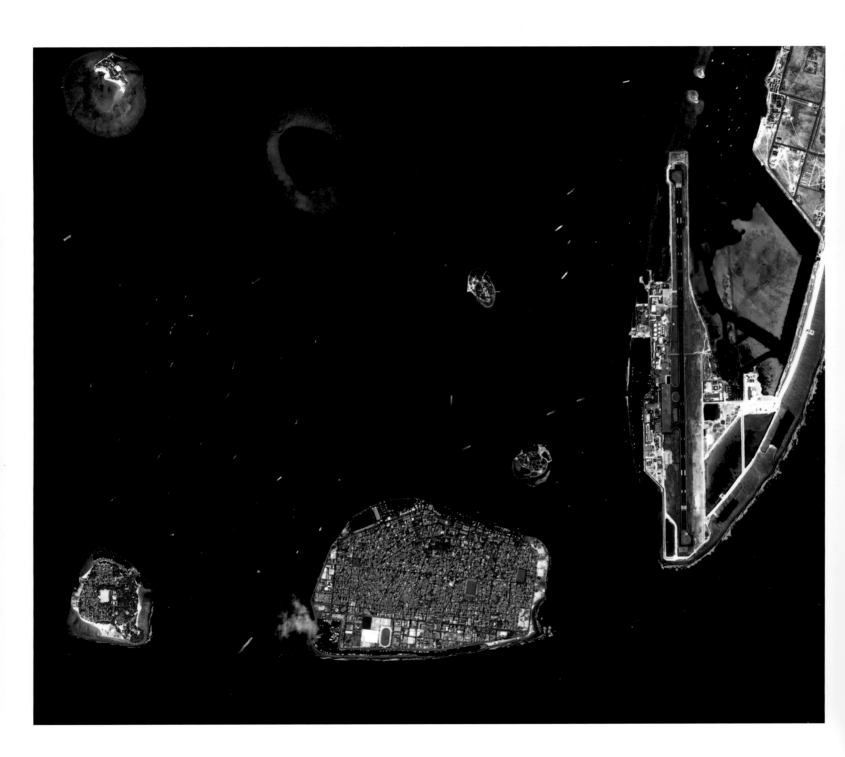

Malé, Maldives
OrbView-3
1:32,000

With an average elevation of only about 2.4 meters, the urban island of Malé is endangered by rising sea levels. The Republic of Maldives was the first country to sign the Kyoto Protocol committing to reduce greenhouse gas emissions that lead to global warming. The Velana International Airport on the separate Hulhulé Island is clearly visible in the right of the image, as is boat traffic essential to the city's commerce.

Iqaluit, Nunavut, Canada
Landsat OLI/TIRS
1:500,000

Iqaluit, meaning "place of many fish," is the only city in the Nunavut territory of Canada. For a portion of every year, snow and ice (light blue) make it inaccessible by road, ship, or railway. Due to its coastal location, it is a traditional fishing area of the Inuit.

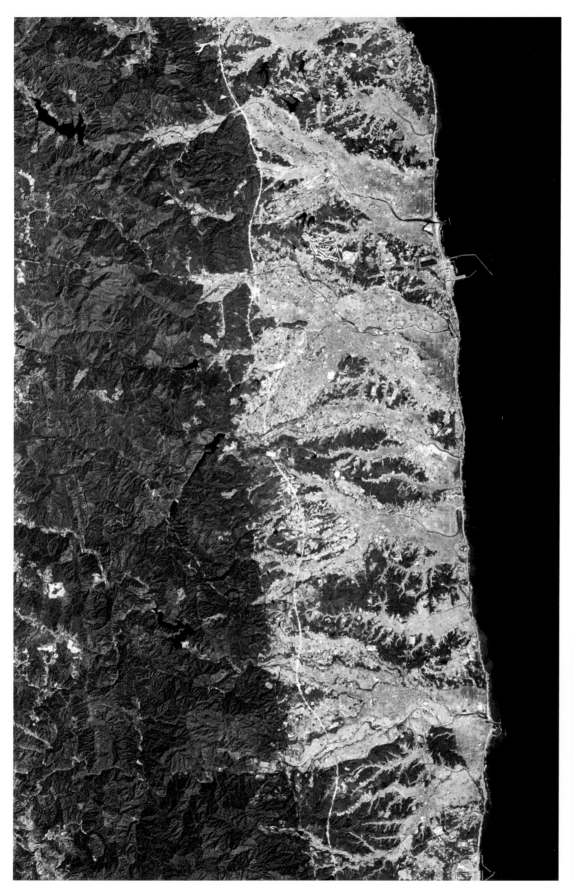

Minamisoma, Japan
ASTER
1:130,000

April 7, 2009

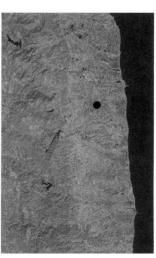

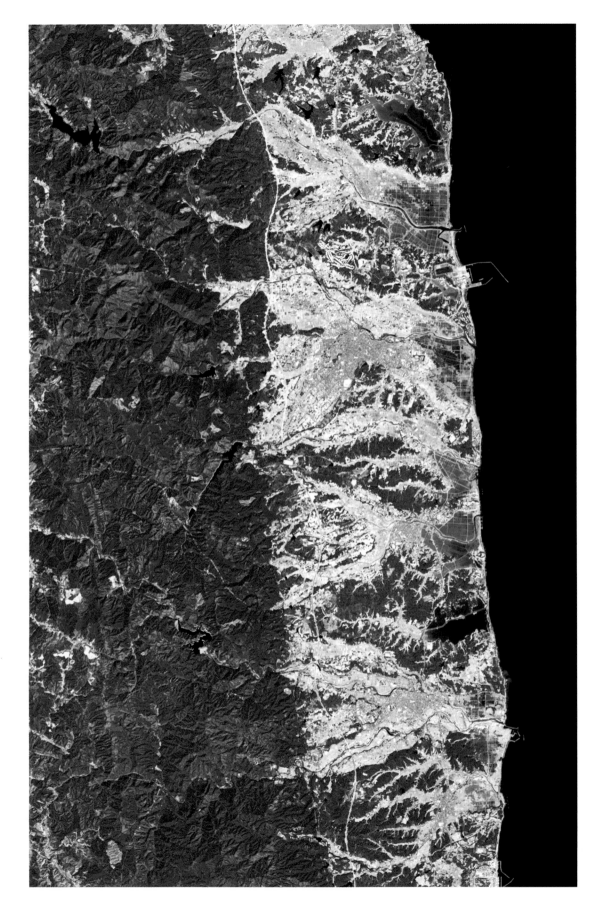

Minamisoma, Japan
ASTER
1:130,000

April 6, 2011

On March 11, 2011, as a result of the Tōhoku earthquake, a tsunami struck the coast of Japan, inundating the coastline and leading to the meltdowns of three reactors at the Fukushima Daiichi Nuclear Power Plant complex (south of pictured region). Standing water can be seen in the April 6, 2011, image pictured in brownish blue along the coast. The original coastal infrastructure can be seen in the April 7, 2009, image.

Christchurch, New Zealand
ASTER
1:295,000

January 15, 2011

On February 22, 2011, a shallow earthquake with an epicenter about six miles (ten kilometers) southeast of Christchurch shook the city, leading to extensive damage throughout the region. Christchurch is pictured here just a month before the earthquake. The Banks Peninsula fills the southeast quadrant of the image. It was formed by volcanic activity in the Miocene period (between eleven million and eight million years ago). The two harbors on the peninsula—Lyttelton (north) and Akaroa (south)—mark the two main active craters of the volcanos that formed the peninsula.

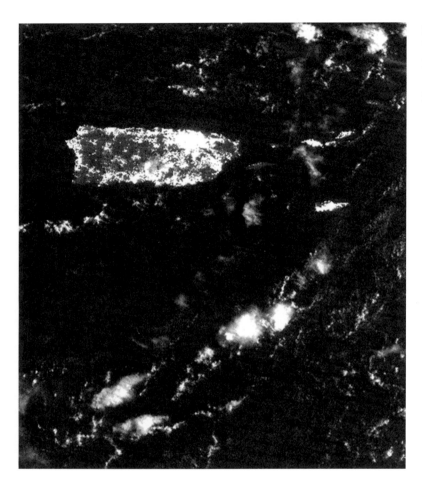

San Juan, Puerto Rico, United States
VIIRS
1:3,900,000

September 21, 2016

San Juan, Puerto Rico, United States
VIIRS
1:3,900,000

September 22, 2016

These images capture illumination on the earth's surface at night. Here, we see the effects of a power plant fire in Puerto Rico that caused massive power outages across the island. San Juan is clearly visible in the northeast of the island. San Juan remains illuminated in both images, indicating that the power plant fire did not affect the main energy supply to this region of the capital city. In 2017, Hurricane Maria left most of Puerto Rico without power.

Pripyat, Ukraine
Landsat TM
1:225,000

April 29, 1986

Pripyat, Ukraine
Landsat OLI/TIRS
1:225,000

October 31, 2015

Two images of Pripyat, Ukraine—a city built to house workers and families of the Chernobyl nuclear power plant—are shown here about thirty years apart. The 1986 image was taken just days after the reactor number 4 meltdown on April 26, 1986. The city's 49,000 residents were evacuated abruptly thirty-six hours later on April 27, 1986, and have never returned due to high radiation levels throughout the city. Agricultural plots have since been overtaken by wild vegetation, as visible in the second image, from 2015, which lacks the characteristic patchwork cropland pattern.

epilogue

History is replete with technological inventions that changed how societies view the world and subsequently understand it. The microscope, the telescope, and the camera fundamentally transformed our ability to see more than with the naked eye, be it up close, far away, or by stopping motion. We are in a similar transformative time in human history. The widespread availability of remote sensing data from satellites orbiting the earth has allowed us to see our planet from new perspectives and enabled new understandings of the relationship between cities and the environment.

This perspective could not have come at a more opportune moment. Contemporary large-scale and rapid urbanization is transforming the planet, and the availability of big data from Earth-observing satellites is transforming how we study these changes. Many of the most pressing sustainable development and environmental challenges have an essential urban component for their solution.

How can we build and operate cities while preserving the natural environments in which they coexist? How can we develop innovative solutions to adapt to climate change, create inclusive communities, and foster cultural and economic development that accounts for the incredible variability within and between urban areas? How can cities mitigate greenhouse gas emissions and be part of the climate change solution? How can we reach the underserved and urban poor who do not have access to transport, clean drinking water, improved sanitation, and electricity? Where can we continue to build urban settlements while saving land for nature, wildlife, and food production? How can cities be more resilient to natural hazards and a changing environment? How will a dynamic earth affect an urbanizing humanity?

The answers to these questions will require new scientific understandings about urbanization and cities. We need this knowledge not for one or even a dozen cities, but for all urban areas worldwide—small or large, developing or established, industrialized or industrializing. Remote sensing imagery can be a central element in developing this new urban science.

Answering these questions also requires engagement at all levels—better local knowledge of cities and neighborhoods, better bridges between the science and policy communities, and a more resilient, connected global community. Ancient cities and towns were planned and managed by a few individuals with a vision for a place. Few cities today are built from scratch on a single vision. In many areas, urbanization is often a piecemeal process that involves many voices and ideas. We all have a role in making urban areas vibrant places.

These images show our impact on the planet— our similarities, differences, challenges, and opportunities. They highlight lessons from the past. We hope they make clear our power to shape the future.

further reading

Yang, Xiaojun, ed. *Urban Remote Sensing: Monitoring, Synthesis and Modeling in the Urban Environment.* John Wiley & Sons, 2011.

Campbell, James B., and Randolph H. Wynne. *Introduction to Remote Sensing.* Guilford Press, 2011.

A Brief History of Landsat: https://en.wikipedia.org/wiki/Landsat_program

Spiro, Kostof. *The City Shaped: Urban Patterns and Meanings Through History.* Bulfinch. New York, 1991.

Campanella, Thomas J., and Witold Rybczynski. *Cities from the Sky: An Aerial Portrait of America.* Princeton Architectural Press, 2001.

Grant, Benjamin. *Overview: A New Perspective of Earth.* Amphoto Books, 2016.

acknowledgments

We are grateful for many people who helped make this volume a reality. This book would have never happened without the patience, guidance, and wisdom of our editor, Joe Calamia. His unwavering support for the project, together with his insights and suggestions, helped to create an immeasurably better end result.

We would also like to thank everyone at Yale University Press who assisted with the book's production: our manuscript editors Joyce Ippolito and Ann-Marie Imbornoni, Christina Coffin and Maureen Noonan our production staff, Eva Skewes our editorial assistant, our rights manager Amy Hawkins, and Nancy Ovedovitz, who gave us early design feedback.

Four people were central to the project. Katie Weber's technical prowess, from image processing and script writing, to building and managing the entire image database, kept the project moving at a clip. The artistic vision of Santiago Villota Cortes was instrumental for creating the bold and lively aesthetics of the project and made the images leap off the pages. Katie McConnell's eye for detail added clarity to the final images and the text. Nicole Westfall's meticulous research on each city was essential for the image descriptions.

This book emerged from discussions years ago with Nick Allen, whose experience and knowledge of cities helped shape our vision of the book. Cary Simmons's creative aesthetic was also central in the initial phases of the project. Ellie Killiam tested and processed numerous images when the book was still just an idea.

We are enormously thankful to the entire Seto Lab at Yale—Kellie Stokes, Chris Shughrue, Kangning Huang, and Bhartendu Pandey—for their honest and critical feedback on all aspects of the project, from image processing to city selection.

We thank Barabara Schoeberl for her work and expertise crafting multiple figures in the remote sensing chapter. We would also like to thank Lindsey Voskowsky for her design and typesetting insight that unified the visual elements of the book.

Caroline Acheatel, Michael Cox, Greg Duncan, Yanin Kramsky, Jessica Leung, Rachel Lowenthal, Maggie Tsang, and Katherine Cooper provided invaluable feedback on the images, themes, and design.

We thank Digital Global for donating images of Palmanova, Zaatari, Eixample, Mysore, and McMurdo station and Miguel Román at NASA Goddard Space Flight Center for providing the Suomi-NPP VIIRS nighttime images of Puerto Rico. We are also grateful for support from Garik Gutman and NASA LCLUC grant NNX15AD43G.

Finally, this project wouldn't have been possible without the unconditional support of our families: Rob Davis, Jing Seto, and Marie and Jim Reba.

index